145
Advances in Polymer Science

Editorial Board:
A. Abe · A.-C. Albertsson · H.-J. Cantow · K. Dušek
S. Edwards · H. Höcker · J. F. Joanny · H.-H. Kausch
T. Kobayashi · K.-S. Lee · J. E. McGrath
L. Monnerie · S. I. Stupp · U. W. Suter
E. L. Thomas · G. Wegner · R. J. Young

Springer

*Berlin
Heidelberg
New York
Barcelona
Hong Kong
London
Milan
Paris
Singapore
Tokyo*

Radical Polymerisation
Polyelectrolytes

With contributions by
I. Capek, J. Hernández-Barajas,
D. Hunkeler, J.L. Reddinger,
J.R. Reynolds, C. Wandrey

Springer

This series presents critical reviews of the present and future trends in polymer and biopolymer science including chemistry, physical chemistry, physics and materials science. It is addressed to all scientists at universities and in industry who wish to keep abreast of advances in the topics covered.

As a rule, contributions are specially commissioned. The editors and publishers will, however, always be pleased to receive suggestions and supplementary information. Papers are accepted for „Advances in Polymer Science" in English.

In references Advances in Polymer Science is abbreviated Adv. Polym. Sci. and is cited as a journal.

Springer WWW home page: http://www.springer.de

ISSN 0065-3195
ISBN 3-540-65210-8
Springer-Verlag Berlin Heidelberg New York

Library of Congress Catalog Card Number 61642

This work is subject to copyright. All rights are reserved, whether the whole or part of the material is concerned, specifically the rights of translation, reprinting, re-use of illustrations, recitation, broadcasting, reproduction on microfilms or in other ways, and storage in data banks. Duplication of this publication or parts thereof is only permitted under the provisions of the German Copyright Law of September 9, 1965, in its current version, and permission for use must always be obtained from Springer-Verlag. Violations are liable for prosecution under the German Copyright Law.

© Springer-Verlag Berlin Heidelberg 1999
Printed in Germany

The use of registered names, trademarks, etc. in this publication does not imply, even in the absence of a specific statement, that such names are exempt from the relevant protective laws and regulations and therefore free for general use.

Typesetting: Data conversion by MEDIO, Berlin
Cover: E. Kirchner, Heidelberg
SPIN: 10691421   02/3020 - 5 4 3 2 1 0 - Printed on acid-free paper

Editorial Board

Prof. Akihiro Abe
Department of Industrial Chemistry
Tokyo Institute of Polytechnics
1583 Iiyama, Atsugi-shi 243-02, Japan
E-mail: aabe@chem.t-kougei.ac.jp

Prof. Ann-Christine Albertsson
Department of Polymer Technology
The Royal Institute of Technolgy
S-10044 Stockholm, Sweden
E-mail: aila@polymer.kth.se

Prof. Hans-Joachim Cantow
Freiburger Materialforschungszentrum
Stefan Meier-Str. 21
D-79104 Freiburg i. Br., FRG
E-mail: cantow@fmf.uni-freiburg.de

Prof. Karel Dušek
Institute of Macromolecular Chemistry, Czech
Academy of Sciences of the Czech Republic
Heyrovský Sq. 2
16206 Prague 6, Czech Republic
E-mail: office@imc.cas.cz

Prof. Sam Edwards
Department of Physics
Cavendish Laboratory
University of Cambridge
Madingley Road
Cambridge CB3 OHE, UK
E-mail: sfe11@phy.cam.ac.uk

Prof. Hartwig Höcker
Lehrstuhl für Textilchemie
und Makromolekulare Chemie
RWTH Aachen
Veltmanplatz 8
D-52062 Aachen, FRG
E-mail: 100732.1557@compuserve.com

Prof. Jean-François Joanny
Institute Charles Sadron
6, rue Boussingault
F-67083 Strasbourg Cedex, France
E-mail: joanny@europe.u-strasbg.fr

Prof. Hans-Henning Kausch
Laboratoire de Polymères
École Polytechnique Fédérale
de Lausanne, MX-D Ecublens
CH-1015 Lausanne, Switzerland
E-mail: hans-henning.kausch@lp.dmx.epfl.ch

Prof. Takashi Kobayashi
Institute for Chemical Research
Kyoto University
Uji, Kyoto 611, Japan
E-mail: kobayash@eels.kuicr.kyoto-u.ac.jp

Prof. Kwang-Sup Lee
Department of Macromolecular Science
Hannam University
Teajon 300-791, Korea
E-mail: kslee@eve.hannam.ac.kr

Prof. James E. McGrath
Polymer Materials and Interfaces Laboratories
Virginia Polytechnic and State University
2111 Hahn Hall
Blacksbourg
Virginia 24061-0344, USA
E-mail: jmcgrath@chemserver.chem.vt.edu

Prof. Lucien Monnerie
École Supérieure de Physique et de Chimie
Industrielles
Laboratoire de Physico-Chimie
Structurale et Macromoléculaire
10, rue Vauquelin
75231 Paris Cedex 05, France
E-mail: lucien.monnerie@espci.fr

Prof. Samuel I. Stupp
Department of Materials Science
and Engineering
University of Illinois at Urbana-Champaign
1304 West Green Street
Urbana, IL 61801, USA
E-mail: s-stupp@uiuc.edu

Prof. Ulrich W. Suter
Department of Materials
Institute of Polymers
ETZ,CNB E92
CH-8092 Zürich, Switzerland
E-mail: suter@ifp.mat.ethz.ch

Prof. Edwin L. Thomas
Room 13-5094
Materials Science and Engineering
Massachusetts Institute of Technology
Cambridge, MA 02139, USA
E-mail. thomas@uzi.mit.edu

Prof. Gerhard Wegner
Max-Planck-Institut für Polymerforschung
Ackermannweg 10
Postfach 3148
D-55128 Mainz, FRG
E-mail: wegner@mpip-mainz.mpg.de

Prof. Robert J. Young
Manchester Materials Science Centre
University of Manchester and UMIST
Grosvenor Street
Manchester M1 7HS, UK
E-mail: robert.young@umist.ac.uk

Contents

Radical Polymerization of Polyoxyethylene Macromonomers in Disperse Systems
I. Capek 1

Molecular Engineering of π-Conjugated Polymers
J.L. Reddinger, J.R. Reynolds 57

Diallyldimethylammonium Chloride and its Polymers
C. Wandrey, J. Hernández-Barajas, D. Hunkeler 123

Author Index Volumes 101–145 183

Subject Index 193

Radical Polymerization of Polyoxyethylene Macromonomers in Disperse Systems

Ignác Capek

Polymer Institute, Slovak Academy of Sciences, Dúbravská cesta 9, 842 36 Bratislava, Slovakia
E-mail: upolign@savba.sk

Dedicated to Professor J.Barton (Polymer Institute, Slovak Academy of Sciences, Dúbravská cesta 9, 842 36 Bratislava, Slovakia) on the occasion of his retirement and 65th birthday.

Polymerization in disperse systems is a technique which allows one to prepare ultrafine and microsize latex particles, as well, and random, comb-like, comb-like, star-like, and graft copolymers. This article presents a review of the current literature in the field of the surfactant-free dispersion or emulsifier-free emulsion polymerization and copolymerization of the polyoxyethylene unsaturated macromonomers. The key factor for the preparation of polymer dispersion is the type of emulsifier and its concentration. When conventional surfactants are used, the high amount of stabilizer is needed to prepare a fine polymer dispersion. Conventional surfactants are held on the particle surface by the physical factors. An interesting alternative arises with the use of reactive surfactants which contain a polymerizable group. The reactive surfactants are incorporated into the polymer matrix or the particle surface layer which prevents them from subsequent migration. Together with a short introduction into some kinetic aspects of radical polymerization of traditional monomers in dispersion, emulsion, miniemulsion and microemulsion, we focus mainly on the organized aggregation of amphiphilic polyoxyethylene macromonomers and radical copolymerization of polyoxyethylene macromonomers with styrene and alkyl (meth)acrylates. We discuss mechanisms of particle growth, particle nucleation, the growth and termination polymer chains, and colloidal stability. Effects of initiator, macromonomer, diluent, continuous phase type and concentration of initiator, macromonomer and additives, the surface activity of macromonomer, the type of organized association of macromonomer or graft copolymer molecules on the polymerization and particle size are evaluated. Variation of molecular weight with the reaction conditions is also discussed.

Keywords. Radical polymerization and copolymerization, Graft copolymer, Polyoxyethylene macromonomers, Organized aggregation of macromonomers

List of Abbreviations and Symbols		2
1	Introduction	5
2	Kinetics of Radical Polymerization of Conventional Monomer and Amphiphilic Macromonomers in Disperse Systems	7
2.1	Dispersion Polymerization	7
2.2	Emulsion Polymerization	13
2.3	Miniemulsions	16
2.4	Microemulsions	17

3	**Micelles of PEO Amphiphilic Macromonomers**	19
3.1	Introduction	19
3.2	PEO Unsaturated Macromonomers	21
3.3	PEO Saturated Macromonomers	24
3.4	PEO Block and Graft Copolymers	25

4	**Dispersion Polymerization of PEO Macromonomers**	27
4.1	Copolymerization of PEO Macromonomers with Styrene	27
4.2	Copolymerization of PEO Macromonomers with Alkyl Acrylates and Methacrylates	33

5	**Emulsion Polymerization of PEO Macromonomers**	34
5.1	Homopolymerization of PEO Macromonomers	34
5.2	Copolymerization of PEO Macromonomers with Styrene	39
5.3	Copolymerization of PEO Macromonomers with Other Comonomers	45

6	**Polymerization of PEO Macromonomers in Other Disperse Systems**	48

7	**Conclusion**	50

8	**References**	52

List of Abbreviations and Symbols

A	acrylic group
A_2	second virial coefficient
AA	acrylic acid
AVA	4,4'-azobis(4-cyanovaleric acide)
AIBN	2,2'-azobiisobutyronitrile
BA	butyl acrylate
BzMA	benzyl methacrylate
BMA	butyl methacrylate
CAC	critical association concentration
C_w	concentration of monomer in water
C_p	concentration of polymer
CMC	critical micelle concentration
CFC	critical flocculation concentration
CFT	critical flocculation temperature
(CL)	chain length

C_1	methyl
tC_4	t-butyl
C_s	chain transfer constant to stabilizer
C_{ss}	chain transfer to solvent
C_{SP}	chain transfer constant for transfer to polymeric stabilizer
D	particle diameter
DLS	dynamic light scattering
D_{50}	volume median diameter
D_f	final particle diameter
DBP	dibenzoyl peroxide
DP_n	number average degree of polymerization
D_w	diffusion coefficient of the radical in water
E_o	overall activation energy
E_p	activation energy for propagation
E_t	activation energy for termination
E_d	activation energy for decomposition of initiator
EO	ethylene oxide unit
f	initiator efficiency
f_w	monomer feed composition
G_a	graft available
G_r	graft required
HLB	hydrofile-lipophile balance
HUFT	homogeneous nucleation model of emulsion polymerization
I	initiator
ISP	inverse suspension polymerization
k_a	second-order radical entry rate coefficient
k_{des}	exit (desorption) rate constant
k_1	a pseudo-zero-order rate of dead chain generation
k_2	a diffusion-controlled rate constant for coalescence of similar-sized particles
k_d	initiator decomposition constant
k_p	propagation rate constant
k_t	termination rate constant
K_{eq}	partition koefficient
k_{tr}	chain transfer rate constant
KPS	potassium peroxodisulfate
SLS	static light scattering
L-PEO	long PEO chains
m	average number of aggregated monomers in one micelle
M_1	macromonomer
M_2	small (traditional) monomer
$[M]_{eq}$	equilibrium monomer concentration
$[M]$	monomer concentration
$[M]_D$	monomer concentration in the diluent
MA	methacrylic acid

MMA	methyl methacrylate
MW_s	molar weight of the monomer
M_n	number-average molecular weight
M_w	weight average molecular weight
MWD	molecular weight distribution (M_w/M_n)
N_p	number of polymer particles
N_o	number of particles containing zero radicals
N_1	number of particles containing one radical
N_D	number of monomer droplets
NBA	N,N-methylenebisacrylamide
ñ	average number of radicals per particle
n	number of units (e.g. EO)
n_M^o	initial number of moles of monomer present per unit volume of water in the reactor
N_A	Avogadro constant
[P]	the concentration of polymeric stabilizer
PEO	poly(ethylene oxide)
PEO-A	acryloyl-terminated PEO
PEO-MA	methacryloyl-terminated PEO
PEO-VB	vinylbenzyl-terminated PEO
PEO-MAL	maleic-terminated PEO
PSt	polystyrene
PMMA	poly(methyl methacrylate)
PVPo	polyvinylpyrrolidone
PVP	poly(vinylpyridine)
Q_{min}	minimum graft coverage required to stabilize particles against same-size coalescence
Q_{max}	maximum graft coverage
r	particle radius
r_s	monomer-swollen particle radius
R_p	rate of polymerization
$R_{p,max}$	maximum rate of polymerization
r_1 and r_2	reactivity ratios
$1/r_2$	relative reactivity of macromonomer M_1
R_h	hydrodynamic radius
R_g	radius of gyration
R	alkyl group
[S]	concentration of stabilizer
SDS	sodium dodecyl sulfate
St	styrene
S-short	short PEO chains
S_{crit}	critical surface area occupied by a PEO chain
t	reaction time
T_g	glass transition temperature
THF	tetrahydrofurane

V_p	volume fraction of polymer particles in the system
VA(50)	2,2' azobis(2-aminopropane) dichloride
VAc	vinyl acetate
W_{1o}	initial weight of macromonomer
W_{2o}	initial weight of monomer
x	fractional conversion
x_1	fractional conversion of macromonomer
α	monomer partitioning coefficient between polymer and diluent
$α_n$	degree of neutralization
ρ	overall radical entry rate constant
$ρ_A$	rate of adsorption of oligomeric radicals by polymer particles
$ρ_D$	first-order radical entry rate coefficient
$ρ_P$	density of the polymer
[η]	intrinsic viscosity

1
Introduction

In this review we summarize and discuss the amphiphilic properties of polyoxyethylene (PEO) macromonomers and PEO graft copolymer molecules, the aggregation of amphiphilic PEO macromonomers into micelles, the effect of organized aggregation of macromonomers on the polymerization process, and the kinetics of radical polymerization and copolymerization of PEO macromonomer in disperse (dispersion, emulsion, miniemulsion, microemulsion, etc.) systems [1–5].

In order to generate stable polymer colloid dispersions in aqueous media, it is necessary to provide a repulsive interaction that outweighs the van der Waals attraction between the particles. This can be achieved in several different ways. First, by electrostatic stabilization, in which the Coulombic repulsion between the charged colloidal particles is operative (for some emulsion systems). Second, by steric stabilization, whereby stability is imparted by nonionic polymers or stabilizers adsorbed or grafted on the particle polymer surface (for dispersion systems). Last, by using a combination of electrostatic and steric stabilization mechanisms, i.e., by electrosteric stabilization. Polymer dispersions stabilized by electrostatic stabilization become unstable at high electrolyte concentrations, in various pH regions, in freeze-thaw cycling, and at high rates of shear. Under these conditions sterically stabilized dispersions can be used.

When conventional surfactants are used in emulsion polymerization, difficulties are encountered which are inherent in their use. Conventional surfactants are held on the particle surface by physical forces; thus adsorption/desorption equilibria always exist, which may not be desirable. They can interfere with adhesion to a substrate and may be leached out upon contact with water. Surfactant migration affects film formation and their lateral motion during particle-particle interactions can cause destabilization of the colloidal dispersion.

An interesting alternative arises with the use of macromonomers (polymerizable surfactants). Amphiphilic PEO macromonomers, the PEO graft copolymers, and oligomers present all the typical properties of conventional nonionic surfactants, such as micelle formation and interfacial tension reduction. In addition, "reactive surfactants" contain a polymerizable group; thus they can overcome some of the difficulties encountered with conventional surfactants and can function not only as surfactants, but can also be incorporated into the surface layer of the latex particles by copolymerization with comonomers. In this manner these reactive surfactants are prevented from subsequent migration; i.e., they cannot be desorbed from the particle surface or excluded from a film. In addition to these advantages, they can confer stability over a wide pH range and at low temperatures depending on the surfactants structure.

The reaction mechanism of amphiphilic macromonomer in disperse systems is a complex function of partitioning of the macromonomer between the different phases, the macromonomer and initiator type, the nature of continuous phase and reaction conditions. The use of amphiphilic macromonomers seems to be of interest in this regard since, due to their hydrophilic-hydrophobic character, they tend to be located mostly at the particle surface favoring its polymerization there. The dispersion and free-emulsifier emulsion polymerization and copolymerization of amphiphilic macromonomers are considered to be very important for the preparation of the graft and comb-like copolymers and non-traditional (functional) polymer latexes. Since the functional oligomers (graft copolymers) are normally used as a minor constituent, it is desirable that a large fraction become incorporated at the particle surface. However, the highly water soluble macromonomers give, by polymerization in water, a large amount of water soluble polymers. Furthermore, the surface yield of the functional group, i.e., the relative amount incorporated at the particle surface, is often low.

Macromonomers are macromolecules with a polymerizable group. The primary factor affecting the polymerization process is the type of macromonomer unsaturated group [3–5]. The reactivity of macromonomers mostly follows the reactivity of the low molecular weight monomers carrying the same functional group. The polymerization run of macromonomer(s) is characterized with a high viscosity of the reaction medium, the low concentration of the unsaturated groups, and high segmental density around both the propagating radical end and the unsaturated group of the macromonomer. These features enhance the diffusion controlled effects on the polymerization kinetic parameters.

Macromonomers afford a powerful means of designing a vast variety of well-defined graft copolymers. These species are particularly useful in the field of polymer blends as compatibilizers and/or stabilizers (surfactants). When macromonomer itself is an amphiphilic polymer, then its polymerization in water usually occurs rapidly as a result of organization into micelles. In copolymerizations, important factors for macromonomer reactivity are the thermodynamic repulsion or incompatibility between the macromonomer and the trunk polymer and its partitioning between the continuous phase and the polymer particles [4, 5].

We begin by describing the current understanding of the kinetics of polymerization of classical unsaturated monomers and macromonomers in the disperse systems. In particular, we note the importance of diffusion-controlled reactions of such monomers at high conversions, the nucleation mechanism of particle formation, and the kinetics and kinetic models for radical polymerization in disperse systems.

2
Kinetics of Radical Polymerization of Conventional Monomer and Amphiphilic Macromonomers in Disperse Systems

The radical polymerization in disperse systems may be divided into several types according to the nature of continuous phase and the polymerization loci: the dispersion, emulsion, miniemulsion, microemulsion, suspension, etc.

2.1
Dispersion Polymerization

Dispersion polymerization was used for preparation of nano- and micron-sized polymer particles and graft copolymers. It is defined as a heterogeneous polymerization by which latex particles are formed in the presence of a suitable steric stabilizer from an initially homogeneous reaction mixture. In fact, dispersion polymerization can be regarded as a special case of precipitation polymerization in which flocculation is prevented and particle size controlled by stabilizer. The solvent selected as the reaction medium must be a good solvent for both the monomer and the steric stabilizer but a poor solvent or a non-solvent for the polymer being formed [1, 2]. The steric stabilizer is used to produce a colloidally stable polymer dispersion. In the absence of this stabilizer, the polymerization produces large particles (agglomerates), and it is known as precipitation polymerization.

In the dispersion polymerization of unsaturated monomers there are two polymerization loci, namely the monomer-swollen polymer particles and the continuous phase. The reasons that the rate of polymerization and the molecular weight of the final polymers prepared by the dispersion polymerizations are larger than in the solution polymerization are due to the compartmentalization of the reaction loci. On the other hand, the rate of polymerization and the molecular weight of the final polymers prepared by the dispersion polymerization are lower than in the conventional emulsion polymerization which may be attributed to the low monomer concentration in the polymer particles, low particle concentration, and very high average number of oligomeric radicals per particle (pseudo bulk kinetics). The relative high number of (oligomeric) radicals existing in the monomer-swollen particles – is due to the large particle size, high radical flux per particle, and high viscosity within the particles. The limiting amount of monomer in the polymer particles is therefore shared by a large number of radicals, resulting in restricted growth events (lower rates and molecular weights for the final polymers).

Generally, the maximum rate of dispersion polymerization is found at low or medium conversions and the rate decreases with conversion. No constant concentration of monomer in polymer particles during the dispersion polymerization can be expected. At the beginning of polymerization, a dispersion polymerization proceeds at a rate similar to a corresponding solution polymerization. After the nucleation period the rate of polymerization increases strongly as a result of a certain compartmentalization of reaction loci, the gel effect, and many growing radicals per particle. It is known that the number of radicals per particle is a function of particle size. When the particle size is larger than 100 nm, two or more radicals can coexist in a particle without instantaneous termination. Thus the gel effect can lead to the Smith-Ewart 3 kinetics (the average number of radicals per particle >>1, see emulsion polymerization). A consequence of these findings is that the rate of polymerization is independent of or slightly dependent on the particle number, but depends instead on the particle volume [1]:

$$R_p = k_p [M]_p (\rho_A V_p / k_t)^{0.5} = \alpha k_p [M]_d (R_i V_p / k_t)^{0.5} \tag{1}$$

where $[M]_p$ is the monomer concentration in particles, $[M]_d$ is the monomer concentration in the diluent, R_i is the rate of initiation, α is the monomer partition coefficient between polymer and diluent (the continuous phase), ρ_A is the rate of adsorption of oligomeric radicals by polymer particles, V_p is the volume fraction of polymer particles at a given time, and k_t is the rate constant of termination. The rate of adsorption of oligomeric radicals by the particles, ρ_A, which determines the average number of radicals per particle, thus plays an important role in determining the polymerization rate [1]. If the heterogeneous polymerization mechanism determines the reaction process the rate of polymerization is only slightly dependent on the initiator or stabilizer concentration.

Several methodologies for preparation of monodisperse polymer particles are known [1]. Among them, dispersion polymerization in polar media has often been used because of the versatility and simplicity of the process. So far, the dispersion polymerizations and copolymerizations of hydrophobic classical monomers such as styrene (St), methyl methacrylate (MMA), etc., have been extensively investigated, in which the kinetic, molecular weight and colloidal parameters could be controlled by reaction conditions [6]. The preparation of monodisperse polymer particles in the range 1–20 µm is particularly challenging because it is just between the limits of particle size of conventional emulsion polymerization (100–700 nm) and suspension polymerization (20–1000 µm).

Dispersion polymerization in polar media is an alternative route to prepare monodisperse polymer particles in the 1–20 µm size range. This is a simple and efficient one-step method. During the copolymerization the polymer precipitates from an initially homogeneous reaction mixture containing monomer, initiator, steric stabilizer, and solvents. Under favorable conditions, monodisperse polymer particles stabilized by a steric barrier of dissolved polymer are formed. The main requirements for the formation of monodisperse polymer particles using this process are (1) the nucleation period should be very short, (2) all the oligomeric radicals which are generated in the continuous phase during the polymerization should be captured by the existing particles before they precipitate and form new particles, and (3) the coalescence between premature particles in the particle growth stage should be prevented.

Polymeric steric stabilizer such as poly(vinylpyrrolidone) (PVPo), poly(acrylic acid), poly(hydroxypropyl)cellulose, etc., are used to prepare monodisperse polymer in dispersion polymerization of monomers such as alkyl acrylates and methacrylates, and styrene in polar media. AB and ABA block copolymers are a second type of steric stabilizer which can be used in dispersion polymerization. For example, the poly(styrene-b-ethylene oxide) was recently used by Winnik et al. [6] in the dispersion polymerization of styrene in methanol.

The nucleation mechanism of dispersion polymerization of low molecular weight monomers in the presence of classical stabilizers was investigated in detail by several groups [2, 6, 7]. It was, for example, reported that the particle size increased with increasing amount of water in the continuous phase (water/ethanol), the final latex radius in their dispersion system being inversely proportional to the solubility parameter of the medium [8]. In contrast, Paine et al.[7] reported that the final particle diameter showed a maximum when Hansen polarity and the hydrogen-bonding term in the solubility parameter were close to those of steric stabilizer.

The nucleation mechanism of amphiphilic macromonomers is well described by the homogeneous nucleation [9, 10], according to which the most important point in the reaction is the instant at which colloidally stabilized particles are formed. After this point, coagulation between similar-sized particles no longer occurs, and the number of particles present in the reaction is constant (see the HUFT theory below). The dispersion polymerization and copolymerization of amphiphilic (hydrophilic) PEO macromonomer is considered to proceed as follows (see Scheme 1): Before polymerization, the low-molecular weight (traditional) monomer (=), macromonomer (---), and initiator (o--o) dissolve completely in the solvent. The oligomer graft copolymers are all produced by polymerization in the continuous phase, accompanied by the decomposition of the initiator. The solubility of these polymers is a function of their molecular weight and the composition of graft copolymers. Polymers with a molecular weight larger than a certain critical value precipitate and begin to coagulate to form unstable polymer particles. The small (primary) particles coagulate on contact, and the coagulation between them is continuous until sterically stabilized particles form. This point is referred to as the critical point, and it occurs if the polymer particles contain sufficient surface active chains at the surface to provide colloidal stability.

The graft copolymers were already used for preparation and stabilization of polymer particles by Barrett [1]. He synthesized a poly(12-hydrostearic acid) macromonomer with a methacrylate end group. This macromonomer was copolymerized with MMA to obtain a preformed comb-graft copolymer, which was successfully used as stabilizer in nonaqueous dispersions of MMA.

Scheme 1. Mechanism of dispersin polymerization (explanation in the text)

Thus the use amphiphilic macromonomers is another method to achieve the particle formation and their subsequent stabilization. Macromonomers can be pre-reacted to form graft copolymers, which are be introduced into the reaction medium afterwards. Macromonomers can also be copolymerized with classical monomers in situ to form graft copolymers. This is a simple and flexible method for producing monodisperse micron-sized polymer particles. Macromonomers can produce ion-free acrylic lattices with superior stability and film forming properties compared to conventional charge stabilized lattices. These non-conventional monomers have been recognized as being useful for the preparation of monodisperse polymer particles. Monodisperse polymer particles in the micron range have found wide applications in various fields. Because of the simplicity of the process for preparation of micron-size monodisperse polymer particles, dispersion (co)polymerization of macromonomers in polar media has recently received great attention.

The fundamental relations for the particle size (D_{50}) and number (N_p), and the fractional conversion (x) or the reaction time (t) in the dispersion polymerization polymerizarion according to the model of Paine [11] (originally derived from the homogeneous nucleation theory [9,10]) are as follow:

$$D^3_{50}=(0.386(CL)6MW_s k_1 k_2 t^2)/(\pi \rho_p N_A)= \qquad (2)$$

$$=(0.386 \times 6MW_s[M]x^2 k_2/\pi \rho_p N_A)(k_t/fk_d[I])^{1/2} \qquad (3)$$

$$N_p=(N_A k_p/0.386 k_2 x)(fk_d[I]/k_t)^{1/2} \qquad (4)$$

The critical point occurs when the graft available equals the minimum graft required (Q_{min}):

$$D_{crit}=(6[M]MW_s Q_{min}/\rho N_A C_s[S]) \qquad (5)$$

$$x_{crit}=d_{crit}^{3/2}(0.386\pi \rho_p N_A k_p/6[M]MW_s k_2)^{1/2}(fk_d[I]/k_t)^{1/4} \quad (x<<1) \qquad (6)$$

where D_{50} is the volume median diameter, D_f is the final particle size, MW_s is the molar weight of the monomer, (CL) is the chain length, C_s is the chain transfer constant to stabilizer, [S] is the concentration of stabilizer, ρ_p is the density of the polymer, N_A is Avogadro's number, k_1 is a pseudo-zero-order rate of dead chain generation, k_2 is a diffusion-controlled rate constant for coalescence of similar-sized particles, f is the initiation efficiency, k_d is the initiator decomposition rate constant, and [I] is the initiator concentration.

This model was used in dispersion polymerization to predict the size of polymer particles stabilized through grafting on hydrophilic polymers such as PVPo. It provides a reasonable description of, for example, PVPo-stabilized polymerization of styrene in polar solvents. The present model does not apply to other types of dispersion polymerization where grafted comb or block copolymer stabilizers are active. The key controlling parameters in this model are the availability of graft and the minimum and maximum coverage, Q_{min} and Q_{max}.

Kawaguchi et al. [12] have modified Paine's model for the dispersion copolymerization of amphiphilic PEO macromonomers. The authors have modeled the variation of the particle size and its distribution with reaction conditions. For example, the expressions for the critical conversion (x_{crit}), the particle radius (r), and the surface area (S) occupied by a PEO chain are as follows:

$$x_{crit}=(r_{crit})^{3/2}(4/3\pi \rho_p N_A k_p/0.386 k_2 W_{20})^{1/2}(2fk_d[I]/k_t)^{1/4} \qquad (7)$$

$$r=x^{1/3}(3W_{20}/\rho_p N_A)^{2/3}(M_D r_1/W_{10} S_{crit})^{1/2}(0.386 k_2/4\pi k_p)^{1/6}(k_t/2fk_d[I])^{1/12} \qquad (8)$$

$$S = x^{1/3}(3W_{20}/\rho_A N_A)^{1/3}(M_D S_{crit}/W_{10}r_1)^{1/2}$$
$$(4\pi k_p/0.386k_2)^{1/6}(2fk_d[I]/k_t)^{1/12}(x/x_1) \qquad (9)$$

where W_{20} is the weight of comonomer (a low-molecular weight monomer) in the reactants, W_{10} the weight of macromonomer in the reactants, x_1 the conversion of macromonomer, M_D the molecular weight of macromonomer, r_1 is the reactivity ratio in copolymerization of monomer (M_2) with macromonomer (M_1) and at the critical point these values are x_{crit}, S_{crit}, r_{crit}, x_{Dcrit}. For example, S_{crit} was written as $S_{crit}=5/3\pi\langle S^2\rangle^{1/2}$ and $\langle S^2\rangle_z^{1/2}$ (for PEO in methanol) is $0.16M_D^{0.585}$.

This model shows that the radius of polymer particle follows simple scaling relationships with the key parameters in the system: $x^{1/3}$, $[\text{comonomer}]_o^{2/3}$, $[\text{macromonomer}]_o^{1/2}$, and $[\text{initiator}]_o^{-1/2}$, where $[\ ]_o$ means initial concentration. These equations also predict that the particle size and stabilization are determined by the magnitude of r_1. In addition the surface area occupied by a hydrophilic (PEO) chain follows $x^{-1/3}$ in the case of azeotropic copolymerization, $x=x_1$. This means that the PEO chain conformation for chains grafted onto the polymer particles change with grafting density.

Lacroix-Desmazes and Guyot [13] applied Paine's model to the dispersion copolymerization of amphiphilic macromonomers and re-discussed this model in terms of possible incorporation of a new parameter – the chain transfer parameter (C_{ss} the chain transfer constant for transfer to solvent-alcohol). The relations for the rate of dead chains (k_1) and chain length (CL) are as follows:

$$k_1 = \{2fk_d[I]/(1+a)\} + C_{ss}k_p[S](2fk_d[I]/k_t)^{0.5} \qquad (10)$$

$$CL = \{2fk_d[I]k_t)^{0.5}/((1+a)k_p[M]) + C_{ss}[Ss]/[M]\}^{-1} \qquad (11)$$

where [M] is monomer concentration, [Ss] is concentration of alcohol, and a is the coefficient depending on the mode of termination (0<a<1).

The critical diameter is reached when G_a (the graft available)=G_r (the graft required):

$$D_{crit} = D_{50} = (6[M]MW_s Q_{min})/(\rho_P C_{SP}[P]N_A) \qquad (12)$$

$$G_a = C_{SP}[P] \times N_A \text{ and } G_r = N_p \pi D_{50}^3/Q_{min} = 6 \times [M]MW_s Q_{min}/\rho_P D_{50} \qquad (13)$$

where C_{SP} is the chain transfer constant for transfer to polymeric stabilizer and [P] is the concentration of polymeric stabilizer.

The critical conversion is given by the following equation:

$$x_{crit} = D^{1.5}_{crit} k_p (\pi \rho_P N_A f k_d[I])/((0.386(CL)3MW_s k_1 k_2 k_t) \qquad (14)$$

The critical particle diameter and the final particle size for copolymerization with macromonomer were re-written by Guyot et al. as shown:

$$D_{crit} = D_{50} = (6[M_1]MW_s Q_{min} r_1)/(\rho_P[M_2]N_A) \qquad (15)$$

$$D_f = (6MW_s/\rho_p N_A)^{2/3}(Q_{min}[M_1]r_1/[M_2])^{0.5}(1/k_p)^{1/3}$$
$$(0.386(CL)k_1 k_2 k_t/2\pi f k_d[I])^{1/6} \tag{16}$$

2.2
Emulsion Polymerization

In emulsion polymerization the compartmentalization of reaction loci and the location of monomer in polymer particles favor the growth and slow down termination events. The contribution of solution polymerization in the continuous phase is strongly restricted due to the location of monomer in the monomer droplets and/or polymer particles. This gives rise to greatly different characteristics of polymer formation in latex particles from those in bulk or solution polymerization. In emulsion polymerization, where polymer and monomer are mutually soluble, the polymerization locus is the whole particle. If the monomer and polymer are partly mutually soluble, the particle/water interfacial region is the polymerization locus.

The emulsion polymerization system consists of three phases: an aqueous phase (containing initiator, emulsifier, and some monomer), emulsified monomer droplets, the monomer-swollen micelles, and monomer-swollen particles. Water is the most important ingredient of the emulsion polymerization system. It is inert and acts as the locus of initiation (the formation of primary and oligomeric radicals) and the medium of transfer of monomer and emulsifier from monomer droplets or the monomer-swollen particle micelles to particles. An aqueous phase maintains a low viscosity and provides an efficient heat transfer.

The emulsifier provides sites for the particle nucleation and stabilizes growing or the final polymer particles. Even though conventional emulsifiers (anionic, cationic, and nonionic) are commonly used in emulsion polymerization, other non-conventional ones are also used; they include reactive emulsifiers and amphiphilic macromonomers. Reactive emulsifiers and macromonomers, which are surface active emulsifiers with an unsaturated group, are chemically bound to the surface of polymer particles. This strongly reduces the critical amount of emulsifier needed for stabilization of polymer particles, desorption of emulsifier from particles, formation of distinct emulsifier domains during film formation, and water sensitivity of the latex film.

The most commonly used water-soluble initiator is the potassium, ammonium, or sodium salt of peroxodisulfates. Redox initiators (Fe^{2+} salt/peroxodisulfate, etc.) are used for polymerization at low temperatures. Oil-soluble initiators, such as azo compounds, benzoyl peroxides, etc., are also used in emulsion polymerization. They are, however, less efficient than water-soluble peroxodisulfates. This results from the immobilization of oil-soluble initiator in polymer matrix, the cage effect, the induced decomposition of initiator in the particle interior, and the deactivation of radicals during desorption/re-entry events [14, 15].

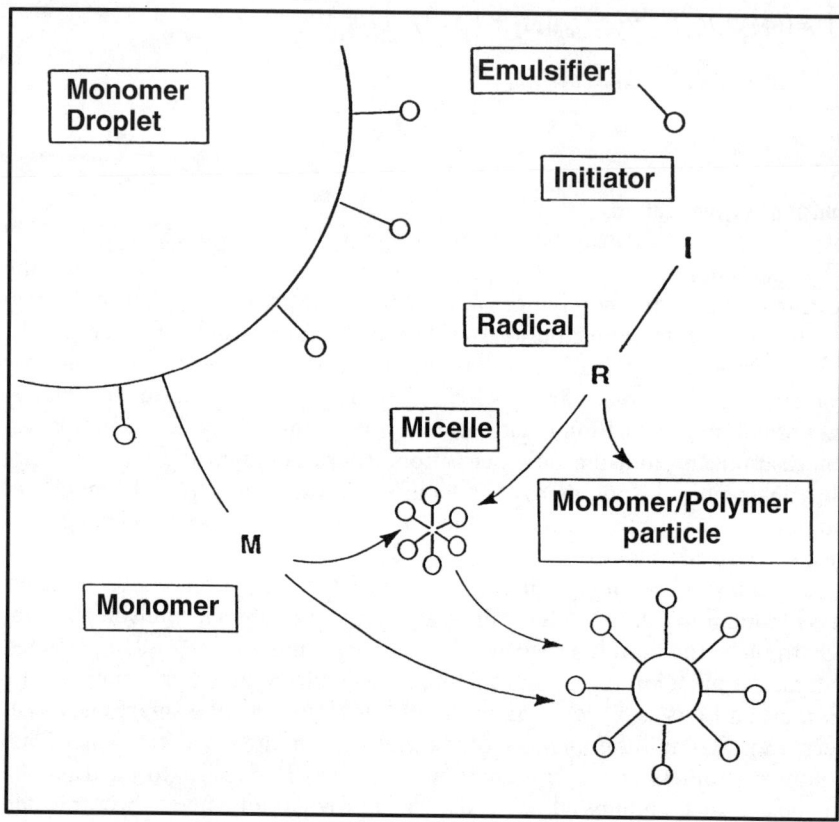

Scheme 2. Mechanism of emulsion polymerization

In accordance with the Smith-Ewart theory, the nucleation of particles takes place solely in the monomer-swollen micelles which are transformed into polymer particles [16]. This mechanism is applicable for hydrophobic (macro)monomers (see Scheme 2). The initiation of emulsion polymerization is a two-step process. It starts in water with the primary free radicals derived from the water-soluble initiator. The second step occurs in the monomer (macromonomer)-swollen micelles by entered oligomeric radicals.

It is usual to consider the course of emulsion polymerization to proceed through three intervals [16, 17]. The particle number increases with time in Interval I, where latex particles are being formed, and then remains constant during Intervals II and II. The monomer concentration in particles is in equilibrium with a monomer saturated aqueous solution. Swelling is limited only by the opposite force of the particle surface/water tension. Hence, the concentration of monomer in the particles is usually taken as constant up to the point where free monomer droplets disappear. In Intervals I and II, the monomer concentration in particles can be regarded to be constant. Interval I ends when the free emulsifier phase disappears. Interval II ends when the monomer droplets disappear. The transition from Interval II to III depends on the monomer/polymer type (the extent of swelling of polymer particles by monomer and the interfacial tension) and the water solubility of the monomer. For vinyl acetate, the transition occurs at ca. 15% conversion (ϕ_m (monomer weight fraction)=0.85) [18], vinyl chloride at ca. 70–80% conversion (ϕ_m=0.3) [18], styrene at ca. 40–50% conversion (water solubility=0.036 wt%), cyclohexyl methacrylate at ca 60% conversion (water solubility=0.016 wt%) [19].

In the case of more water-soluble monomers and (amphiphilic) macromonomers, the Smith-Ewart [16] expression does not satisfactorily describe the particle nucleation. The HUFT [9, 10] theory, however, satisfactorily describes the polymerization behavior or the particle nucleation of such unsaturated hydrophilic and amphiphilic monomers. The HUFT approach implies that primary particles are formed in the aqueous phase by precipitation of oligomer radicals above a critical chain length. The basic principals of the HUFT theory is that formation of primary particles will take place up to a point where the rate of formation of radicals in the aqueous phase is equal to the rate of disappearance of radicals by capture of radicals by particles already formed. Stabilization of primary particles in emulsifier-free emulsion polymerization may be achieved if the monomer (or macromonomer) contains surface active groups. Besides, the charged radical fragments of initiator increases the colloidal stability of the polymer particles.

Thus in the emulsifier-free emulsion copolymerization the emulsifier (graft copolymer, etc.) is formed by copolymerization of hydrophobic with hydrophilic monomers in the aqueous phase. The free-emulsifier emulsion polymerization and copolymerization of hydrophilic (amphiphilic) macromonomer and hydrophobic comonomer (such as styrene) proceeds by the homogeneous nucleation mechanism (see Scheme 1). Here the primary particles are formed by precipitation of oligomer radicals above a certain critical chain length. Such primary particles are colloidally unstable, undergoing coagulation with other primary polymer particles or, later, with premature polymer particles and polymerize very slowly.

The polymer particles are formed by association of amphiphilic macromonomers and graft copolymer molecules. The graft copolymers (precipitated oligomers) associate with each other to form organized structures and are also absorbed by the polymer particles. These agglomerates are supposed to consist of a hydrophobic core (rich in hydrophobic comonomer units) and a hydrophilic shell (rich in hydrophilic macromonomer units). As the polymerization proceeds further, the growth events appear mostly in the polymer particles. The water-phase polymerization generates the oligomer radicals which are absorbed by particles (nucleation is suppressed).

The following rate equations are used for most analyses of the emulsion polymerization [9, 10, 16]:

$$R_p = dx/dt = k_p[M]_{eq}\bar{n}N_p/N_A = k_p[M]_{eq}\bar{n}N_p/n_M^o N_A \tag{17}$$

where $[M]_{eq}$ is the equilibrium monomer concentration in the polymer particles, $\bar{n}$ is the average number of free radicals per latex particle, and n_M^o is the initial number of moles of monomer present per unit volume of water in the reactor.

The dependence of final particle number on the initiator or emulsifier concentrations according to the micellar and homogeneous nucleation theories is given by the following equation:

$$N_p \propto [\text{initiator}]^{0.4} [\text{emulsifier}]^{0.6} \tag{18}$$

The time evolution of the average number of radicals per particle, $\bar{n}$, is given by:

$$d\bar{n}/dt = \rho(1-2\bar{n}) - 2k_{des}\bar{n}^2 \tag{19}$$

where ρ is the total entry rate coefficient, and k_{des} is the exit rate coefficient. The value of k_{des} is given by

$$k_{des} = (3D_w k_{tr} C_w)/(r_s^2 k_p C_p) \tag{20}$$

where D_w is the diffusion coefficient of the radical in water, k_{tr} is the rate constant for transfer to monomer, C_w is the concentration of monomer in the aqueous phase, and r_s is the monomer-swollen particle radius.

2.3
Miniemulsions

Miniemulsions are oil-in-water emulsions prepared by using a mixed emulsifier system comprising an emulsifier and a coemulsifier such as a fatty alcohol or a long chain alkane. The two main characteristics of miniemulsions are their good stability and their small droplet size in the submicron range. Hallworth and Carless [20] proposed that the stability of miniemulsions containing long-chain alkanes or fatty alcohols was derived from the existence of an adsorbed emulsifier/alkanol film at the oil/water interface whose rheological properties made collisions between droplets more elastic. On the other hand, Davis and Smith [21] suggested that the stabilizing effect of the long-chain additives occurred as a result of the prediction of Higuchi and Misra [22] on the destabilization of emulsion by molecular diffusion. The high stability of the miniemulsions is probably a result of both effects. The low water solubility of the coemulsifier may prevent a degradation of the miniemulsions due to molecular diffusion. The existence of the condensed emulsifier layer at the oil/water interface, as a result of the enhanced emulsifier adsorption onto the droplets, may prevent the degradation by droplet-droplet coalescence. It is supposed that the coemulsifier reduces the equilibrium monomer concentration in the polymer particles. A coemulsifier such as an amphiphilic (like a fatty alcohol) or hydrophobic (like a long chain alkane) macromonomer is supposed to increase the stability of polymer miniemulsions.

Miniemulsion polymerizations follow a different mechanism from the conventional (macroemulsion) emulsion polymerizations. Radicals generated in the aqueous phase enter emulsified monomer droplets (monomer-swollen micelles) as primary or oligoradicals and propagate to form monomer-swollen polymer particles. Nucleation of monomer droplets leads to a different rate of polymerization and a different final product. A coemulsifier greatly enhances the stability of small monomer droplets and therefore increases particle numbers. Miniemulsions still follow the typical Smith-Ewart kinetic model and are not simply "small" versions of suspension polymerizations [23].

The conversion-time curves appear to be very similar to the shape typical of emulsion polymerization, i.e., an S-shaped curve is attributed to the autoacceleration caused by the gel effect (Smith-Ewart 3 kinetics, $\bar{n} \gg 1$). The rate of polymerization-conversion dependence is described by a curve with two rate maxima. The decrease in the rate after passing through the first maximum is ascribed to the decrease of the monomer concentration in particles. Particle nucleation ends between 40 and 60% conversion, beyond the second rate maximum. This is explained by the presence of coemulsifier which stabilizes the monomer droplets against diffusive degradation.

It is accepted that the radical entry rate coefficient for miniemulsion droplets is substantially lower than for the monomer-swollen particles. This is attributed to a barrier to radical entry into monomer droplets which exists because of the formation of an interface complex of the emulsifier/coemulsifier at the surface of the monomer droplets [24]. The increased radical capture efficiency of particles over monomer droplets is attributed to weakening or elimination of the barrier to radical entry or to monomer diffusion by the presence of polymer. The polymer modifies the particle interface and influences the solubility of emulsifier and coemulsifier in the monomer/polymer phase and the close packing of emulsifier and coemulsifier at the particle surface. Under such conditions the residence time of entered radical increases as well as its propagation efficiency with monomer prior to exit. This increases the rate entry of radicals into particles.

The first mathematical model for nucleation in monomer droplets was proposed by Chamberlain et al. [25]. In this model, polymer particles were considered to be formed only upon the entry of the radicals into the monomer droplets. The rate of particle formation was expressed as a first-order entry process into monomer droplets:

$$dN_p/dt = \rho_D N_D \tag{21}$$

where ρ_D is the first-order entry rate coefficient of free radicals into the monomer droplets, N_D is the number of monomer droplets, and t is the time.

2.4
Microemulsions

Microemulsions (monomer-swollen micellar solution, micellar emulsions, or spontaneous transparent emulsions) are dispersions of oil in water made with emulsifier and coemulsifier molecules. In many respects they are small-scale versions of emulsions. They are homogeneous on a macroscopic scale but heterogeneous on a molecular scale. They consist of oil and water domains which are separated by emulsifier monolayers. Microemulsions are thermodynamically stable, transparent or translucent, and homogeneous systems with a particle size of about 5–50 nm. These media are multicomponent liquids that exhibit long-term stability, have a low apparent viscosity, and are generally isotropic. This may seem at first sight surprising, but it is due to the very low interfacial tensions between oil and water microdomains. When the tension is sufficiently low ($<10^{-3}$ dyne cm^{-1}), it can be shown that surface energy can be compensated for by dispersion entropy, thus minimizing the free energy of the system. Usual emulsifiers cannot lower the interfacial tensions between oil and water to such ultralow values. A coemulsifier, such as a short-chain alcohol, is frequently necessary. The use of macromonomers as a coemulsifier seems to be of great interest due to the incorporation of a large fraction of convenient macromonomer into the particle surface. This is the case if the macromonomer reduces the rigidity of the interface film, allowing the transition from well-organized phase towards an isotropic microemulsion.

It was observed that the titration of a coarse emulsion by a coemulsifier (a macromonomer) leads in some cases to the formation of a transparent microemulsion. Transition from opaque emulsion to transparent solution is spontaneous and well defined. Zero or very low interfacial tension obtained during the redistribution of coemeulsifier plays a major role in the spontaneous formation of microemulsions. Microemulsion formation involves first a large increase in the interface (e.g., a droplet of radius 120 nm will disperse ca. 1800 microdroplets of radius 10 nm – a 12-fold increase in the interfacial area), and second the formation of a mixed emulsifier/coemulsifier film at the oil/water interface, which is responsible for a very low interfacial tension.

The most significant difference between emulsions (opaque) and microemulsions (transparent) lies in the fact that stirring or increasing the emulsifier concentration usually improves the stability of coarse emulsion. This is not the case with microemulsions – their formation depends on specific interactions among the constituent molecules. If these interactions are not realized, neither intensive stirring nor increasing the emulsifier concentration will produce a microemulsion. On the other hand, once the conditions are right, spontaneous formation occurs and little mechanical work is required [26].

Microemulsion polymerizations follow a different mechanism from the conventional emulsion polymerizations. The most probable locus of particle nucleation was suggested to be the microemulsion monomer droplets [27], although homogeneous nucleation was not completely ruled out. The particle generation rate in microemulsion polymerization is given by an expression similar to Eq. (21), which was used for the miniemulsion polymerization of styrene [28]:

$$dN_p/dt = k'_a[R]_w N_D \tag{22}$$

where k'_a is the second-order radical rate coefficient for entry into the microemulsion droplets (monomer-swollen micelles), and $[R]_w$ is the concentration of free radicals in the aqueous phase.

The particle population balance can be expressed as follows (a zero-one condition for the radical distribution, at most one growing radical in a particle):

$$dN_p/dt = k_a[R]_w(N_0-N_1) - k_{des}N_1 + k'_a[R]_w N_D \qquad (23)$$

$$N_p = N_0 - N_1, \ \bar{n} = N_1/N_p$$

where N_0–N_1 are the number of polymer particles containing zero or one radical, respectively, k_a is the radical rate coefficient for entry into the polymer particles, k_{des} is the radical desorption rate coefficient, and $\bar{n}$ is the average number of radicals per particle. The high desorption rate in the microemulsion runs influences the polymerization process. It results from a small size of the microemulsion droplets or particles.

Guo et al. [29] have estimated the entry rate coefficient, k'_a, of radicals into micelles (microemulsion droplets) to be 7×10^5 cm^3 mol^{-1} s^{-1}, which is several orders of magnitude smaller than k_a, the entry rate coefficient into the polymer particles. This was ascribed to the difference of the surface area of microemulsion droplets and polymer particles. The condensed interface layer or the possibly high zeta-potential of the surface of the microemulsion droplets may hinder the entry of radicals.

The presence of a very large number micelles indicates that radicals are captured predominantly by the monomer-swollen micelles. Each entry of a radical to a monomer-swollen micelle leads to a nucleation event and therefore the particle number increases with conversion. The particle growth is supposed to be a result of propagation of monomer and the agglomeration of primary particles. The dead monomer-swollen polymer particles and uninitiated monomer-swollen micelles serve as a reservoir of monomer. Solution or bulk polymerization kinetics seem to govern the microemulsion polymerization process [30, 31].

3
Micelles of PEO Amphiphilic Macromonomers

3.1
Introduction

An emulsifier is a molecule that possesses both polar and nonpolar moieties, i.e., it is amphiphilic. In very dilute water solutions, emulsifiers dissolve and exist as monomers, but when their concentration exceeds a certain minimum, the so-called critical micelle concentration (CMC), they associate spontaneously to form aggregates – micelles. Micelles are responsible for many of the processes such as enhancement of the solubility of organic compounds in water, catalysis of many reactions, alteration of reaction pathways, rates and equilibria, reaction loci for the production of polymers, etc.

The outer-core region of the micelle, commonly referred to as the palisade layer, may provide a medium of intermediate polarity that affects the energetics of transition state formation. The primary influence of micelles is to concentrate all reactants in or near the micelles. Micellar catalysis exploits the ability of surfactants that are virtually water-insoluble. When ionic surfactants are employed, polar or ionic reactants that are freely soluble in water may also be concentrated near the micelles by electrostatic or dipole interactions [32]. The increase in reaction rates depends on a number of factors, including the location of the solubilized reactant in the micelle; nonpolar compounds partition into the micelle core while more polar compounds are formed closer to the micelle-water interface. The extent of solubilization, ionic charge of the micelle, and the size and shape of the micelle are also important factors.

An important group of surface-active nonionic synthetic polymers (nonionic emulsifiers) are ethylene oxide (block) (co)polymers. They have been widely researched and some interesting results on their behavior in water have been obtained [33]. Amphiphilic PEO copolymers are currently of interest in such applications as polymer emulsifiers, rheology modifiers, drug carriers, polymer blend compatibilizers, and phase transfer catalysts. Examples are block copolymers of EO and styrene, graft or block copolymers with PEO branches anchored to a hydrophilic backbone, and star-shaped macromolecules with PEO arms attached to a hydrophobic core. One of the most interesting findings is that some block micelle systems in fact exists in two populations, i.e., a bimodal size distribution.

With nonionic PEO emulsifiers, intermolecular interactions vary with temperature and types of metal ions and solvents. At low temperatures, nonionic emulsifiers are hydrophilic and form normal micelles. At higher temperatures they are lipophilic and form reverse micelles. A weak interaction with metal ions favors the stability of associates against moisture. On the other hand, a strong interaction may lead to a completely amorphous system. Ethanol as a co-solvent is a moderate solvent for PEO at low temperatures, but its power improves as the temperature is raised [34]. This means that solutions of the PEO copolymers in water and ethanol have opposing temperature coefficients of solubility: negative for water and positive for ethanol.

The water solubility of $R-(EO)_n$ types of nonionic emulsifiers is derived from the weak interaction between the ether oxygen of EO unit and water. It was suggested that each EO unit in the PEO chain, requires three molecules of water to form a hydrated complex [35]. This hydrogen bond complex is destroyed if the solution is taken above the melting point of the PEO. Water usually acts as plasticizer when present in hydrophilic PEO polymers and T_g values decrease with increasing water contents [36]. This phenomenon in the PEO-water system is observed up to 1 mol water/ether group. Beyond this a rise in T_g is observed and water acts as an antiplasticizer.

In aqueous solution, amphiphilic molecules aggregate into micelles above the critical micelle concentration. Such solutions have been the object of research for many years, with special interest in shape and size of these micellar aggregates [37]. Size and shape (spherical, wormlike, or disklike micelles) depend strongly on the molecular structure of the amphiphilic molecule.

These amphiphilic compounds consist of a hydrophobic moiety connected to hydrophilic (crystallizable) PEO chains. PEO blocks having fewer than 12–13 EO units are typically noncrystalline at room temperature and are miscible with the hydrophobic portion of the polymer, exhibiting water-like clarity. As the number of EO units in the block increases, the consistency of the emulsifier will change to a hazy opaque liquid, then to a paste, and finally to a solid. PEO chains exhibit a marked increase in viscosity over the molecular weight range, presumably due to their highly polar (strongly interactive) nature. Gel formation and the high viscosity may be noted. The melting point of PEO emulsifier type is directly related to the portion of the PEO hydrophile [38].

Water solubility, among other properties, changes according to hydrophilic content of the nonionic emulsifier. A nominal HLB value for ethoxylated hydrophobes can be calculated by dividing the weight percentage of EO by 5. In essence, the calculated HLB or the polarity index is a direct function of EO content. Higher HLB values are therefore indicative of higher water solubility. Besides, the water content and cloud point may be correlated with HLB values.

3.2
PEO Unsaturated Macromonomers

Poly(ethylene oxide) (PEO) macromonomers constitute a new class of surface active monomers which give, by emulsifier-free emulsion polymerization or copolymerization, stable polymer dispersions and comb-like materials with very interesting properties due to the exceptional properties of ethylene oxide (EO) side chains. They are a basis for a number of various applications which take advantage of the binding properties of PEO [39], its hydrophilic and amphipathic behavior [40], as well as its biocompatibility and non-absorbing character towards proteins [41]. Various types of PEO macromonomers have been proposed and among them the most popular are the acrylates and methacrylates [42].

One of the most important applications involves the formation of effective emulsifiers using macromonomer. Comb polymers with a hydrophilic base and hydrophobic grafted chains have been used for a long time for the steric stabilization of various dispersions, emulsions, etc. [1], but the approach using macromonomer greatly simplifies the formation of such substances and extends the range of the possible structural variants. Using poly(ethylene oxide) macromonomer, various graft copolymers can be prepared [43].

Richtering et al. [44–46] have investigated a nonionic emulsifier (methacryloyl-terminated PEO, $(EO)_8$-C_{11}-MA), i.e., $-CH_2=C(CH_3)COO(CH_2)_{11}(OCH_2CH_2)_8 OCH_3$. The emulsifier (macromonomer) forms small spherical micelles with a molar mass of 44,000 g mol^{-1}, radius of 3.4 nm, an aggregation number of 70, and CMC 3.4×10^{-4} mol dm^{-3}. At high temperatures small micelles aggregate into random clusters. Phase separation on heating, with a lower critical solution temperature of 43 °C and a hexagonal liquid crystalline phase, were observed.

A different behavior of monomeric and polymeric molecules (polymacromonomers) in aqueous solution was found which was discussed in terms of strong correlation between molar mass and solution structure. Amphiphilic molecules try to minimize the contact area of hydrophobic groups with water molecules.

With monomeric molecules, the aggregation number of micelles is determined by equilibrium thermodynamics. In polymeric molecules, however, topological constraints are imposed on the system. If the degree of polymerization exceeds the aggregation number of the monomeric micelle, unsaturated sites of the polymeric molecules become available (directed to the aqueous phase) and intermolecular interactions (agglomeration) occur. In the case of polymer with $M_w = 6.23 \times 10^5$, typical surfactant behavior was found.

The weight-average molecular weight (M_w), intrinsic viscosity ($[\eta]$ cm^{-3} g^{-1}), and hydrodynamic radius (R_h nm^{-1}) increase with increasing temperature. The second virial coefficient (A_2) decreases with temperature: 2.3×10^{-4} mol cm^3 g^{-1} at 10 °C and 0.8×10^{-4} mol cm^3 g^{-1} at 30 °C.

At low temperatures, small spherical micelles are formed which aggregate to random clusters with increasing temperature. No indications for a transition from spherical to anisotropic micelles were found. At higher temperature, the molar mass increases which can be explained by aggregation of spherical micelles to random clusters. The increase of $[\eta]_{red}$ at finite concentration with increasing temperature is a consequence of an increase in hydrodynamic volume due to aggregation. It was estimated that, on average, four water molecules are bound to one ethylene oxide group but this number decreases with temperature.

The dependence of reduced viscosity on concentration is curved upward, which could be interpreted as entanglement of rod-like micelles and/or a strong micellar interaction [47]. The increase in microviscosity results mainly from the growth in micellar size [48].

Kawaguchi et al. [49] have applied the solubility of pyrene to characterize the micelle structure of PEO macromonomers. There are three noteworthy features of pyrene solubility dependence on polymer (PEO) concentration C_p. In the region of low C_p, [pyrene] is nearly constant or increases slightly with C_p, but beyond a certain C_p it increases quite steeply. Sharpness of this transition decreases with the decrease of the alkyl chain length. Above this transition point, the slope increases gradually and becomes constant in the higher C_p region. The differences in slope indicate differences in the partition coefficients between the micelles and solution, α, for pyrene. The α values are quite small, 1–3 orders of magnitude smaller than that in other micellar systems. Thus the hydrophobic domains in the micelles resulting from the present surfactants are not so amenable to dissolution of pyrene, as compared to those of more conventional surfactant micelles such as SDS. The α values increase with increasing alkyl chain length of PEO macromonomers.

The ratio I_1/I_3 (intensities of the first and third bands in the pyrene fluorescence spectrum) changes from 1.8 in water to about 0.6 in aliphatic hydrocarbon solvents and decreases with increasing C_p. The high I_1/I_3 (1.58–1.65) at low C_p (ca. 1.58–1.65) was related to the existence of premicelles.

The microviscosity of PEO micelles is comparable to that of SDS. To explore the possibility of specific interactions between pyrene and the PEO group, H-NMR spectra were measured for the PEO macromonomers (C_1-(EO)$_{57}$-C_7-VB) in the presence and absence of pyrene. The spectra did not show any evidence of interaction with the PEO. On the contrary, it was suggested that PEO interacts strongly with benzene or other aromatic hydrocarbons. The microviscosity of the micelles was also estimated from the pseudo-first-order rate coefficient for excimer formation for micellar pyrene [50].

The aggregation numbers N_{agg} is determined as 27 for C_1-$(EO)_{53}$-C_4-VB and 38 for C_1-$(EO)_{53}$-C_7-VB micelles by analysis of fluorescence curves. A micelle formation mechanism is proposed for nonionic polymeric surfactants with weakly hydrophobic groups. At low concentrations of PEO macromonomers, large loosely aggregated structures involving the PEO chains are formed. At higher concentrations normal micelles form. These are star-shaped, with a hydrophobic core surrounded by a corona of PEO chains.

The polymerizability of R-$(EO)_n$-VB macromonomers has its maximum (R_p) around n=15–20 [51]. This finding was related to the micelle formation which is expected to be unfavored for either too long or too short chain length of PEO. The macromonomers and their polymacromonomers with very short R are soluble in water and therefore they lose their amphiphilic nature. The parameters of R and n of macromonomer (R-$(EO)_n$-VB) were found to correlate with the formation of micelles and their structure. In the aqueous phase the scattering intensity increased with the concentration of macromonomer above the CMC. The critical micellar concentration in water was found to be in the range from 3.3×10^{-5} to 7.1×10^{-5} mol dm^{-3} for several R-$(EO)_n$-VB macromonomers.

The parameter m/V_m was found to be related to the rate of polymerization; it means the average number of the molecules per unit volume of the micelle (the relative density of the micelle), where m is the average number of aggregated monomers in each micelle and V_m the average volume of micelle. According to this finding, some micelle structures of PEO macromonomers were suggested. The macromonomers carrying hydrophobic and hydrophilic end groups agglomerate in such a way that the hydrophobic unsaturated groups concentrate in the core and the hydrophilic PEO groups coil around the core to make the shell. The macromonomers carrying hydrophobic groups at both ends of the hydrophilic PEO chain form a loop-like arrangement of macromolecules in the micelle.

Ferguson et al. [52] showed that for all series of PEO emulsifiers (macromonomers) the CMC increases correspondingly with increasing length of the ethoxylated chain. For example, with increasing EO number in a C_{12-14}-$(EO)_y$-A macromonomer type (A=acrylate group) the CMC (10^{-5} mol dm^{-3}) increased in the series

$$4.5 \ (y=12) < 5.3 \ (14) < 6.7 \ (20) < 13.3 \ (30). \tag{24}$$

This dependence is extremely weak with the series C_{18}-$(EO)_y$ and C_{18}-$(EO)_y$A which have very high nominal ethoxylate lengths (>30). It suggests that the CMC converges to some maximum value in any nonionic series of this type. At this point, the main factor seems to be the length of the hydrophobe. The CMCs of the surfactants with C_{18} hydrophobe are much lower than those with the C_{12-14} hydrocarbon chain, i.e., the CMC for MA-$(EO)_{30}$-C_{18} and MA-$(EO)_{30}$-C_{12-14} is 1.3×10^{-5} and 13.3×10^{-5} mol dm^{-3}, respectively.

A comparison of the data for C_{12-14}-$(EO)_y$ with those of C_{12-14}-$(EO)_y$-A shows a consistent reduction in the CMC. For example, the CMC for C_{12-14}-$(EO)_{30}$-OH and C_{12-14}-$(EO)_{30}$-A is 31.6×10^{-5} and 13.3×10^{-5} mol dm^{-3}, respectively. In general, acrylation may reduce the CMC by about a factor of two.

The cloud point temperature passes through a maximum at ca. 20–30 EO units. For example, with increasing EO number in $(EO)_y$-C_{12-14}, the cloud point (°C) increased in the following series:

$$67 \ (y=2) < 75 \ (4) < 78 \ (10) < 80 \ (20) \tag{25}$$

When macromonomer itself is an amphiphilic polymer, its polymerization in water occurs unusually rapidly as a result of organization into micelles. Cohin et al. [53] have shown that non-initiated monomer-swollen micelles act as a reservoir of monomer and surfactant during the polymerization of macromonomer in the polymer (polymerizing micelles) particles. From light scattering measurements, relatively low critical micellar concentration (CMC) values were obtained together with rather large aggregation numbers, when compared to those for conventional nonionic surfactants. This type of behavior of the PEO macromonomers cannot be interpreted simply in terms of the current theory of micelle formation. There is no explicit reason why this PEO with very small hydrophobic groups (styrene or MA) should aggregate in water. This may be attributed to the intermolecular interactions of PEO chains and their dependence on the continuous phase type and temperature. For traditional surfactants (where the alkyl chain group is much larger than C_5-hydrocarbon), the onset of association (CMC) corresponds to hydrophobic domain formation. This phase transition takes place over a narrow range of concentration where CMC is reached

Short chain amphiphiles can be incorporated into the backbone of the polymer chains. The resulting graft macromolecules are able to form both intrachain and interchain aggregates. Polymeric surfactants assemble into a variety of intrachain micelles. These polymeric surfactants and/or amphiphilic polymacromonomers can also form mixed aggregates which incorporate free monomeric (macromonomer with a very small hydrophobic group) surfactants.

3.3
PEO Saturated Macromonomers

An often studied group of nonionic surfactants (emulsifiers) is poly(ethylene oxide) monoalkyl ether of the general structure: C_mH_{2m+1}-$(OCH_2CH_2)_n$-OH; (C_m-$(EO)_n$).

In aqueous solutions of C_m-$(EO)_n$ amphiphilic molecules, two interesting features are observed. First, isotropic micellar solutions undergo phase separation on heating. Such behavior is typical of hydrophobic interaction and is also observed for several water-soluble polymers. Hydrophobic interaction results from a change of order in the water structure [54]. Second, at high concentration, liquid crystalline phase behavior is observed with several structures [55].

Phase diagrams of many C_m-$(EO)_n$ systems were found to demonstrate the complex influence of hydrophilic-hydrophobic balance on miscibility gap and mesophases [37]. An unambiguous description of the solution structure is difficult because size and shape of micellar aggregates can change with temperature and concentration.

Mixed micelles, comprising both polymerized and free surfactants, are formed once the critical association concentration (CAC) of the free amphiphiles is reached. The CAC is typically much smaller than the CMC for the formation of free micelles. As the fraction of unpolymerized surfactants within the mixed micelles grows, their structure approaches that of free micelles.

3.4
PEO Block and Graft Copolymers

Block or graft copolymers in a selective solvent can form structures due to their amphiphilic nature. Above the critical micelle concentration (CMC), the free energy of the system is lower if the block copolymers associate into micelles rather than remain dispersed as single chains. Often the micelles are spherical, with a compact core of insoluble polymer chains surrounded by a corona of soluble chains (blocks) [56]. Addition of a solvent compatible with the insoluble blocks (chains) and immiscible with the continuous phase leads to the formation of swollen micelles or polymeric microemulsion. The presence of insoluble polymer can be responsible for anomalous micelles.

Several morphological studies of polymeric microemulsion have involved graft and block PSt/PEO copolymers in ternary solvent mixtures containing toluene, water, and either an alcohol or an amine acting as a cosurfactant [57, 58]. The co-emulsifier increases the dispersion capacity of the copolymers.

Trace amounts of water are found in most organic solvents. Water strongly affects the structures formed by PSt-PEO block copolymers and induces formation of microemulsions as well. Unusually large structures are observed in dry solutions that disappear upon addition of a small amount of water. Large aggregates were also observed in solutions of PSt-PEO diblock copolymers in selective solvents for polystyrene and were attributed to the crystallizability of PEO [59]. It was suggested that the crystallizability of PEO is responsible for the large aggregates found in dry cyclopentane; at higher water concentrations, water molecules dissolve the PEO chains, reducing the driving force for the formation of large aggregates. Big aggregates were not observed in methanol, a selective solvent for PEO.

At very small water concentrations, PEO chains are linked together by hydrogen bonds with water acting as a "gluing" agent [60]. The large aggregate size does not change as the temperature is varied from 0 to 50 °C. Thus the strong interactions between hydrophilic moieties (PEO) promote the formation of structures not present at higher water contents.

Cogan and Gast [61] presented dynamic light scattering (DLS) data of polystyrene-polyoxyethylene diblock copolymers in cyclopentane in the presence and absence of water at 23 °C. Since water is a solvent for PEO and cyclopentane is close to a θ solvent for PSt (T_θ=19.5 °C), the micelles consist of a PEO core swollen with water and protected by a polystyrene corona.

In a solution of a larger polymer, PSt/PEO – 1730/170, with a polymer and water content of 1500 and 28 ppm, respectively, micelles with R_h of 34 nm were observed. Since the micelle R_h values measured at 30, 60, and 90 °C were the same within experimental error, it was concluded that the micelles are relatively monodisperse. The micelle hydrodynamic radius increases from 34 to 48 nm as the water content is increased from 28 to 98 ppm in a 1500 ppm solution of the 1730/170 block copolymer.

The smaller block copolymers contain a much higher percentage of PEO and insoluble PSt block. Micelles with R_h of 13 nm in solutions of 5800 ppm of polymer and 155 ppm of water were formed. R_h increases to 18 nm at 1030 ppm of water. It was suggested the single chains have R_h values of the order of 2 nm.

Block copolymers of polystyrene (PSt, hydrophobe) and polyoxyethylene (PEO, hydrophile) form spherical micelles in water when the length of water soluble PEO is significantly larger than that of the insoluble PSt portion of the molecule [62]. In analogy with low molecular weight surfactants, one defines the onset of intermolecular association as the critical micelle concentration (CMC). Theories of polymer micellization predict that the concentration of free, unassociated block copolymers is close to that of the CMC.

The PEO-rich PSt-*b*-PEO block copolymers form spherical micelles in aqueous solutions [63]. The DLS measurements indicate the presence of a bimodal size distribution – two very narrowly distributed species. The smaller more mobile species had R_h corresponding to the star model of block copolymer micelles. However, 99% or more of the block copolymer is present as simple micelles.

Fluorescence techniques have been used with great success in the study of PEO-*b*-PSt micelles [64]. In this study, the effect of polymer concentration on the fluorescence of pyrene present in water at saturation was studied. Three features of the absorption and emission spectra change when micellization occurs. First, the low-energy band of the (S_2-So) transition is shifted from 332.5 to 338 nm. Second, the lifetime of the pyrene fluorescence decay increases from 200 to ca. 350 ns, accompanied by a corresponding increase in the fluorescence quantum yield. Third, the vibrational fine structure changes, as the transfer of pyrene from a polar environment to a nonpolar one suppresses the permissibility of the symmetry-forbidden (0,0) band.

PSt-*b*-PEO block copolymers were suggested to form spherical micelles in water solution with a dense core composed of the insoluble PSt. Thus they possess a core of a pure PSt phase surrounded by a water-swollen corona of PEO chains. One anticipates that the properties of the corona resemble those of typical nonionic PEO-based emulsifiers. The core is much larger for block copolymer micelles leading to larger aggregation numbers. When traces of pyrene are added to these solutions, the pyrene penetrates the core phase. Several spectroscopic properties are changed upon transfer into the more hydrophobic environment.

4
Dispersion Polymerization of PEO Macromonomers

4.1
Copolymerization of PEO Macromonomers with Styrene

The rate of dispersion copolymerization of PEO-MA macromonomer and styrene was found to increase with increasing initiator concentration {VA – water soluble, DBP (dibenzoyl peroxide) – oil soluble, [PEO-MA]=0.06 mol dm^{-3}, [styrene]=2.13 mol dm^{-3}, in ethanol/water, v/v 4/1) [65, 66]}:

$$R_{p,max} \propto [VA]^{0.6} \text{ and } R_{p,max} \propto [DBP]^{0.8} \tag{26}$$

Exponents 0.6 and 0.8 deviate from both the homogeneous nucleation and micellar models. Thus the bimolecular termination between the growing radicals is suppressed. This behavior may also result from the surface activity of the graft copolymer formed. The higher the surface activity of graft copolymer the higher the particle number. This behavior would indicates that the graft copolymer formed within the particles (with DBP) is more efficient.

One of the common approaches to infer the termination mechanism is based on the evaluation of the dependence of the molecular weight on the initiator concentration. The following dependences [67]

$$M_n \propto [DBP]^{-0.79} \text{ and } M_w \propto [DBP]^{-0.82} \tag{27}$$

are in favor of the first-order radical loss process as it was already proposed by the kinetic data analysis (Eq. 26). The molecular weights of PSt-*graft*-PEO copolymers are related to the colloidal stability. The more efficient stabilizer (PSt-*graft*-PEO) was formed in the DBP runs (a larger concentration of polymer particles was observed). This finding is in a good agreement with the kinetic data analysis (Eq. 26). This difference should be discussed in terms of the variation of PEO fraction in the graft copolymer. However, the copolymer composition was nearly the same for both runs. The values of M_w were generally one order of magnitude larger for the VA runs than those for the DBP ones. Thus the graft copolymers with lower molecular weight were active as a stabilizer. The large PSt/PEO graft and block copolymers form district microdomains [68] and/or initiate the interparticle (micelle) interaction (aggregation) [46].

The molecular weights increase with conversion and the increase is much more pronounced for the VA runs. The strong increase of molecular weight with conversion results from the transfer of reaction loci from the solution to the polymer particles. At low conversion the polymers are mostly formed in the continuous phase in which the termination rate is high. In polymer particles the local concentration of monomer is much higher than that in the continuous phase and, therefore, the growth events are favored in the polymer particles.

The molecular weights of polymer products of the DBP runs are relatively low and very similar to those in the solution copolymerization [69]. These results indicate that the monomer-DBP saturated polymer particles act as a "homogeneous" reaction loci in which rapid termination (by primary radicals, induced decomposition of initiator, etc.) dominates.

The dispersion copolymerizations of PEO-MA or PEO-VB with styrene were faster than those in solution [69]. The higher rates of polymerization at higher conversion are ascribed to association of macromonomers and graft copolymer molecules which favors the growth events (increases the concentration of unsaturated groups). The values of the ratio ($k_p/k_t^{0.5}$) for PEO macromonomers are approximately one order of magnitude higher than those obtained for the low molecular weight monomers (St and MMA). Thus the compartmentalization of reaction loci due to organization of amphiphilic macromonomers and their graft copolymers favors the growth events.

The dispersion copolymerization of PEO-MA macromonomer and styrene is presented in Figs. 1 and 2 [70]. The rate-conversion plot is curved with a maximum at very low conversion. In all runs, neither the gel effect nor the stationary interval were observed. The strong decrease of the rate of polymerization with increasing conversion results from a decrease in the monomer concentration at the reaction loci (mainly in the polymer particles). The low monomer concentration in particles is a reason why the gel effect may be operative only at very low conversion.

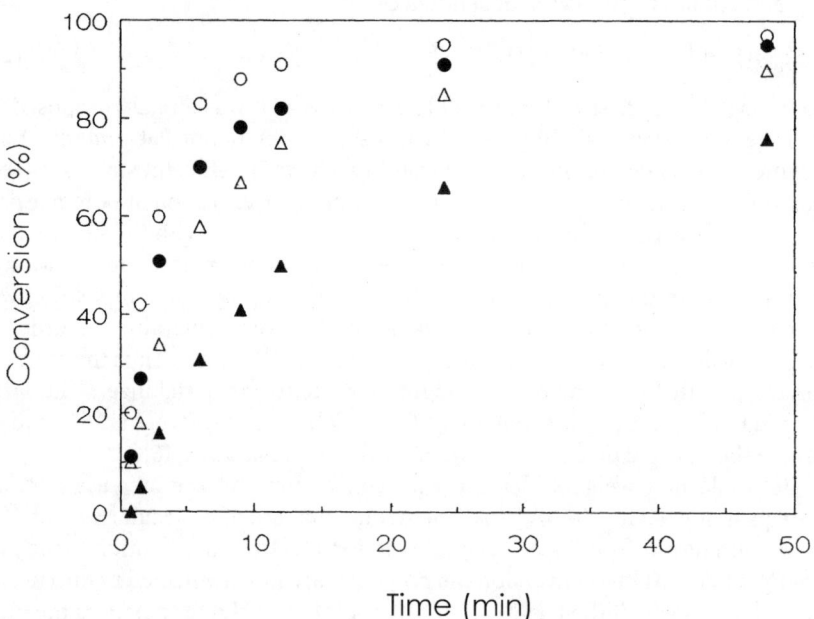

Fig. 1. Dependence of monomer conversion in the the dispersion copolymerization of PEO-MA macromonomer and styrene on reaction time and [PEO-MA] (M_n=100). Recipe: 5 ml of ethanol/water (4/1, v/v), 1.11 g styrene, 0.0273 g VA, temp. 60 °C. [PEO-MA]× 10^2/ (mol dm^{-3})=2 (▲), 4 (△), 6 (●), 8(○)

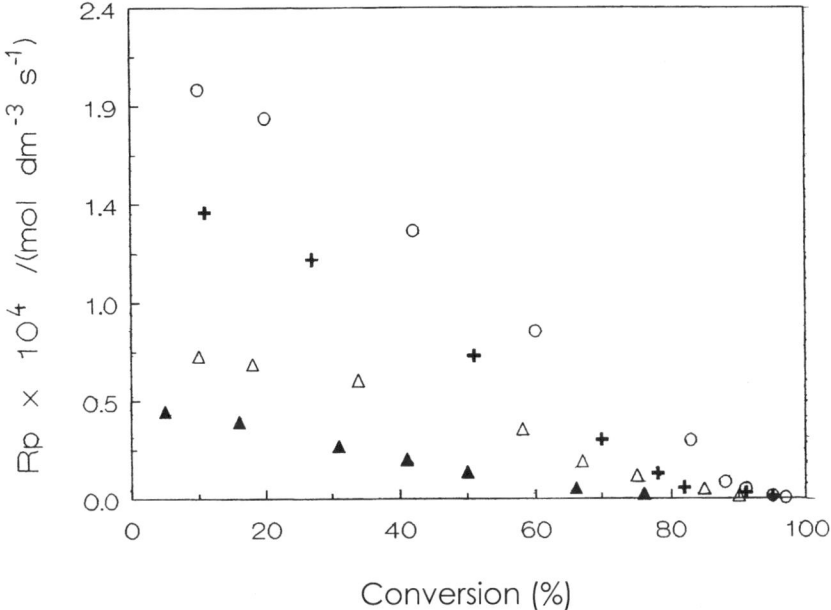

Fig. 2. Dependence of the rate of polymerization in the dispersion copolymerization of PEO-MA macromonomer and styrene on reaction time and [PEO-MA] (M_n=1000). [PEO-MA] × 10^2/(mol dm^{-3})=2 (▲), 4 (△), 6 (✚), 8 (○)

The dependence $R_{p,max}$ vs [PEO-MA] is curved, i.e., the reaction order x increases from 0.5 to 4. This behavior is attributed to the exponential increase of polymer particle number with macromonomer concentration.

The reaction order 4.4 obtained from the plot N_p vs [PEO-MA] is much larger than that for the classical stabilizer or emulsifier (close to 1.0). This finding indicates that the PEO-MA macromonomer (the graft copolymer) is much more efficient than the classical stabilizer. This finding was attributed to several factors: first, the PEO units are chemically bound on the particle surface; second, the fraction of buried PEO units decreases with increasing PEO fraction in the graft copolymer; and third, the high organized aggregation of PEO macromonomer and graft copolymer.

The decrease in M_n of the graft copolymer with increasing [PEO-MA] was attributed to the decrease of monomer concentration in particles and the chain transfer events. The reaction order of [PEO-MA] exponentially decreased with increasing macromonomer concentration.

The dispersion copolymerization of PEO-MA and PEO-VB macromonomers was found to produce monodisperse polymer particles. The PSD (D_w/D_n) was found to be close to 1.08 and it decreased with increasing the initiator and macromonomer concentration but increased with conversion [69, 70].

In copolymerization of styrene, more monodisperse and smaller-size particles were obtained with the PEO-MA macromonomer with a longer alkylene spacer group, thus indicating their effectiveness as reactive surfactant increases in the series [71]

$$C_1\text{-}(EO)_{48}\text{-}(MA) < C_1\text{-}(EO)_{48}\text{-}C_6\text{-}(MA) < C_1\text{-}(EO)_{48}\text{-}C_{10}\text{-}(MA) \tag{28}$$

Thus the more hydrophobically enhanced MA end groups are more effectively grafted into the particle surface. The first one induced some precipitation.

A stable dispersion was also obtained with $C_1\text{-}(EO)_{17}\text{-}C_{10}\text{-}MA$, a macromonomer, and $C_{12}\text{-}(EO)_{19}\text{-}MA$ with a similar HLB produces extensive coagulation almost from the beginning of polymerization, indicating ineffectiveness of the hydrophobic group introduced at the other end of the MA group.

The rate of the dispersion copolymerization of PEO-MA with styrene increased with increasing temperature [72], which was attributed to the increased rate of initiation and particle formation. The overall observed activation energy (E_o) for the dispersion copolymerization of PEO-MA or PEO-VB macromonomer and styrene was found to be 48 or 53 kJ mol^{-1}. These values are much lower than those (ca. 90–100 kJ mol^{-1} [73]) obtained in the solution polymerization of low molecular weight monomers.

The overall activation energy (E_o) is given by the expression

$$E_o = E_p - E_t/2 + E_d/2 \tag{29}$$

where E_p is the activation energy for propagation, E_t is the activation energy for termination, and E_d is the activation energy for decomposition of initiator, a complex function of the monomer and particle concentration. It is known that the polymerization of macromonomers is diffusion controlled, the reactivity of macromonomer growing radicals is low due to the high segment density at the radical end, polymer chains and macromonomers are highly incompatible, composition of polymer particles is heterogeneous, etc. Under such reaction conditions the $E_d/2$ parameter should be constant and the product ($E_p - E_t/2$) is supposed to vary. With increasing temperature the viscosity of particle interior (increased bimolecular termination) decreases which increases the incorporation of PEO groups into the particle interior.

The role of the graft copolymer formed in dispersion polymerization of PEO maleic macromonomers ($R_1OCOCH=CHCO\text{-}(CH_2\text{-}CH_2\text{-}O)_n\text{-}R_2$, PEO-MAL) and styrene was investigated by Lacroix-Desmazes and Guyot [74], where a continuous phase was ethanol/water and the macromonomer (PEO-MAL) type varied as follows: $R_1=H$, $R_2=CH_3$ and $n=49$.

The conversion-time curves for the dispersion polymerization of styrene in the presence of PEO-MAL and PVPo (polyvinylpyrrolidone) showed in both runs that the gel effect was at ca. 20% conversion. It was found that the gel effect was accompanied with a rise of the molecular weight and a broadening of the molecular weight distribution.

The macromonomer system gives a polymer dispersion with larger particles ($D_{n,mac}=1860$ nm, $D_{n,PVP}=980$ nm, 3.1 wt% for styrene), narrower size distribution {D_w/D_n=1.04 (PEO-MAL) and 1.13 (PVPo)} and lower (without) coagulum (3.9 wt% with PVPo) compared with the PVPo system.

It was found that at the end of the polymerization, the incorporation yield of macromonomer is very low, i.e., 3.3%. At this point, however, the macromonomer conversion is more than 90%. Furthermore, a small fraction of the macromonomer is weakly adsorbed, whereas the major part (88.1%) is lost in the serum. This means that only a small part of the macromonomer actually takes part in the stabilization. Thus the chemically bonded as well as the physically adsorbed species stabilize the polymer particles.

The analysis of the reaction serum (the continuous phase without polymer particles) at the end of polymerization led to the conclusion that the molecular weight of the soluble oligomers of styrene and PEO macromonomer varied from 200 to 1100 g mol^{-1}. This indicates that the critical degree of polymerization for precipitation of oligomers in this medium is more than ten styrene units and only one macromonomer unit per copolymer chain. Several reasons for the low molecular weight of the soluble copolymers were proposed, such as the thermodynamic repulsion (or compatibility) between the PEO chain of the macromonomer and the polystyrene macroradical, the occurrence of enhanced termination caused by high radical concentration, and, to a lower extent, a transfer reaction to ethanol [75].

The colloidal stable polymer dispersions, the monodisperse polymer particles, and high conversions (85–100%) can be obtained with most of the other macromonomers (MAL, VB, and MA) of PEO ($M_{w,PEO}$=2000)) [76]. Also, when macromonomers are used (3.1 wt% based on styrene), there is practically no coagulum produced. This is not the case in the presence of polymerizable PEO surfactants (surfmer I: R_1=CH$_3$(CH$_2$)$_{11}$-, R_2=H, n=34 and surfmer II: R_1=CH$_3$(CH$_2$)$_{11}$-, R_2=H, n=42) despite the higher amounts of stabilizer used (up to 60 wt% of coagulum). Furthermore, the particles are more monodisperse with PEO macromonomer (D_w/D_n=1.025 for PEO-MA and 1.13 for PVPo) compared to those with surfmer. Comparatively poorer results were obtained with conventional surfactants such as ethoxylated nonylphenol, even when used in large amounts.

The incorporation yields of the stabilizers are always very limited, but the highest value is obtained for the PEO-MA macromonomer, and the lowest for the amphiphilic compound (surfmer) with the longest hydrophobic sequence ca. 0.5%:

Grafted%/PEO-type: 3.3/MAL<5.7/MAI<11.6/VB<14.7/MA (30)

The mechanism of polystyrene and poly(methyl methacrylate) particle formation in the presence of PEO-MA macromonomer in the presence of conventional stabilizer (PVPo) and the graft copolymers (PSt-*graft*-PEO), respectively, was discussed [77]. At the beginning of dispersion polymerization (in methanol) of MMA (0–250 s) using PVPo, very small particles were formed (12–35 nm in diameter). The population of bigger particles was roughly stabilized at ca. 345 nm in diameter. In the dispersion polymerization of styrene, small particles

(12 nm) were detected, and the average particle size of the bigger population was stabilized at ca. 92 nm. Using PEO-MA (M_w=2000) macromonomer, small particles of about 11 nm were sparingly detected, and the average diameter of bigger population was stabilized at ca. 88 nm, which is in the same range as for the PSt particles using PVPo as stabilizer. The aggregation of these unsaturated nuclei is suggested to be very fast, leading to particle greater than 300 nm and 80 nm for PMMA and PSt, respectively, within a few seconds. The measured average particle size is quickly stabilized at a value which corresponds to the diameter of the primary (premature) particles.

The estimated data (using Paine's model) were found to be in good agreement with experimental measurements obtained in St/PVPo and St/PEO-MA systems. However, in the case of MMA/PVPo, the model (e.g., D_f – the final particle diameter) deviates strongly from experimental data. The deviations were discussed in terms of partitioning of the stabilizer (conventional, grafted stabilizer) between the particle surface and the serum and the efficiency of the graft stabilizer which depends on:

1. relative molecular weight of the grafts: the higher the molecular weight of the grafts, the higher the probability of adsorption on the particles;
2. solubility parameter of the serum: the greater the difference in the solubility between the grafts and the serum the higher the probability of adsorption on the particle;
3. compatibility between the stabilizing polymer and the core polymer: the lower the compatibility between the stabilizing polymer and the core polymer the lower the risk of burying the graft stabilizer;
4. proximity of the polymerization temperature from the glass transition temperature T_g of the particles; the higher the T_g of the polymer particles the lower the possibility of the stabilizing polymer chains to reorient on the surface of the particles.

The preparation of monodisperse micronsized polymer particles by radical dispersion copolymerization of styrene with PEO macromonomers having *p*-styrylbutyl endgroup (M_n=2000) in methanol/water (90/10, v/v) and modeling of colloidal parameters were shown in [12, 78]. The polymer particle diameter varied within the range 0.3–0.5 µm. The dependence of the particle size on the experimental conditions was modeled into the expression

$$r \propto x^{0.33} W_{10}^{2/3} (r_1/W_{20} S_{crit})^{0.5} \tag{31}$$

This model is based on the particle formation during polymerization where the polymer particles are sterically stabilized by graft-copolymerized PEO chains on the particle surface. In the later stage the polymer particles were supposed to grow in size mainly by copolymerization of monomers occluded in the particles which may favor the substrate monomer (styrene) over the macromonomer as compared to the composition in the continuous phase.

It was estimated that for a PEO segment of about 50 units, the area which can be covered by the random coil, i.e., of unperturbated dimensions, is about 1100 Å^2. This value is not far from data obtained for most stabilizers except for

PEO-VB macromonomer. On the other hand, if one takes into account the adsorbed species, the values are much lower, and closer to the saturation values (around 159 Å^2) where the conformation of the PEO chains should be a thick brush.

The stable polymer dispersions with small-sized polymer particles of diameter ≥60 nm were prepared by dispersion copolymerization of PEO-MA macromonomer with styrene, 2-ethylhexyl acrylate, acrylic and methacrylic acids, and butadiene at 60 °C [79]. The particle size was reported to decrease with increasing macromonomer fraction in the comonomer feed. Besides, it varied with the type of the classical monomer as a comonomer. T_g of polymer product was found to be a function of the copolymer composition, the weight ratio macromonomer/monomer, and monomer type and varied from 50.6 to 220.4 °C.

4.2
Copolymerization of PEO Macromonomers with Alkyl Acrylates and Methacrylates

High conversions (close to 100%) can be obtained by the dispersion copolymerization of PEO-MA with butyl acrylate initiated by a water-soluble initiator (VA) [80]. The conversion curves have a shape similar to that for the dispersion copolymerization of PEO-MA with styrene. In runs with AIBN the final conversion was around 90% and/or the polymerization was very slow at high conversion.

The dependence of R_p on conversion was described by a curve with a maximum at ca. 20–30% conversion. The initial increase in the rate of polymerization was ascribed to the increase in the particle concentration, the gel effect, and the monomer saturation conditions. The decrease in the rate above ca. 30% conversion results from the decrease of the concentration of monomer in the polymer particles. The particle size was observed to increase while the number of particles decreased with conversion. The decrease of particle number with conversion probably results from the agglomeration of sticky polyBA particles and/or the intermolecular (interparticle) interaction of PEO graft polymers. The reaction order x=0.7 ($R_{p,max}$ vs [AIBN]^x) supports the contribution of the first radical loss process. This behavior, however, is also contributed to by the variation of the surface activity with the molecular weight of graft copolymer, i.e., the lower molecular weight of graft copolymer the higher the number of polymer particles (the lower surface activity of the graft copolymer) (see above).

It was reported that with increasing [PEO-VB] ((C_1-(EO)$_{53}$-alkyl-VB, where alkyl chains were C_1, C_4 and C_7)), the latex particle radius decreased [12] as

$$r \propto [PEO\text{-}VB]^{-0.56} \tag{32}$$

On the other hand, with decreasing alkyl chain length separating the styrene (vinylbenzene) group from PEO the particle size increases. This was interpreted in terms of differences in r_1 (a reactivity parameter) values. It was assumed that the apparent reactivity of the macromonomer ($1/r_1$) (characterizing the relative reactivity of the macromonomer towards poly(butyl methacrylate) radical) in the polar medium increases with the length of the alkyl chain. Macromonomers with long alkyl chains are more liable to be adsorbed onto non-sterically stabilized parts of the hydrophobic latex particle surface. This would increase both the local concentration of macromonomer and the particle stability.

The longer the PEO chain the smaller the particles at the same macromonomer concentration:

$$r \propto (n_{EO})^{-0.59} \tag{33}$$

This result corresponds to an increase in the surface area occupied by a single PEO chain. The particle size was found to increase strongly with increasing butyl methacrylate (BMA) concentration, but to slightly decrease with increasing initiator concentration as follows:

$$r \propto [BMA]^{0.82}[AIBN]^{-0.1} \tag{34}$$

The reaction order 0.82 was attributed to decreasing S_{crit} and the low fraction of PEO in a graft copolymer. Copolymerized macromonomer may not be adsorbed onto the particle surface but may transform into the particle interior and/or the continuous phase [74, 76]. A slight decrease in the particle size with [AIBN] may be attributed to the nucleation of larger numbers of primary particles due to the formation of graft copolymers with higher stabilizing efficiency (graft copolymers with lower molecular-weight are more efficient as stabilizers [11]).

Dispersion copolymerization of PEO-MA macromonomers (C_1-$(EO)_{48}$-MA, C_1-$(EO)_{48}$-C_6-MA, C_1-$(EO)_{48}$-C_{10}-MA) with MMA was successful in producing very stable PMMA dispersions of micron size [81]. In this case, however, C_1-$(EO)_{48}$-MA was more effective in giving monodisperse particles than C_1-$(EO)_{48}$-C_{10}-MA (the reverse is true with styrene, see above). The particles obtained were found to have uneven surfaces with a number of craters. These results suggest that some compatibility between PMMA and PEO chains and also between PMMA and the medium (methanol/water) may play a role in controlling the particle formation.

5
Emulsion Polymerization of PEO Macromonomers

5.1
Homopolymerization of PEO Macromonomers

Gramain and Frere [82] observed that the free radical polymerization of ω-methacryloyl terminated PEO macromonomers in the aqueous phase deviates from the solution polymerization. Polymerizations initiated by KPS in water were much faster than those that proceeded in the solution. Low molecular weight polymers were formed in the aqueous systems (ca. up to 20 macromonomer units were incorporated into polymer molecules).

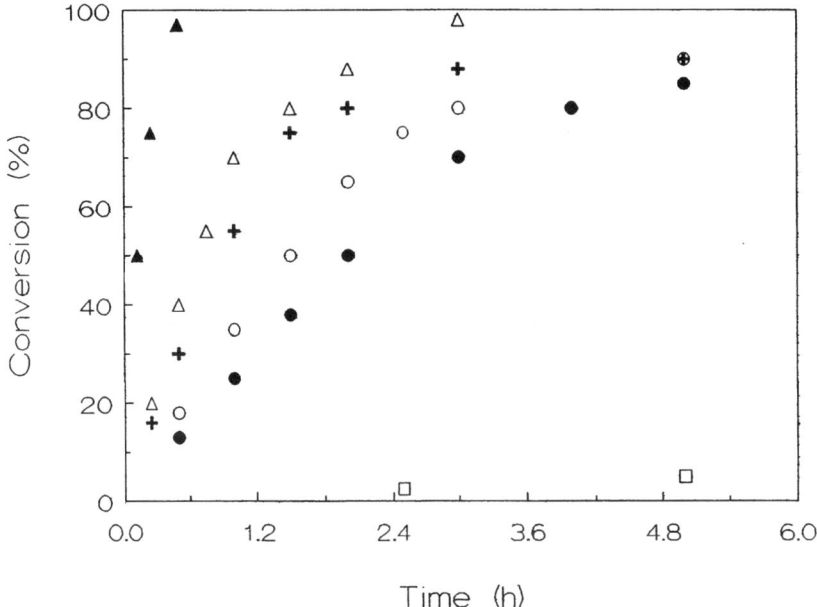

Fig. 3. Dependence of monomer conversion in the emulsifier-free emulsion polymerization of PEO-VB macromonomers on the reaction time and the PEO-VB type [81]. Recipe: [PEO-VB]=0.045 mol dm^{-3}, [AVA]=0.45×10^{-3} mol dm^{-3} (in water), [AIBN]=2.25 ×10^{-3} mol dm^{-3} (in benzene), 60 °C. In water: C_{18}-(EO)$_{35}$-VB (▲), C_8-(EO)$_{43}$-VB (△), C_1-(EO)$_{25}$-VB (✚), C_4-(EO)$_{39}$-VB (○), tC_4-(EO)$_{32}$-VB (●), in benzene: C_4-(EO)$_{39}$-VB (□)

The mechanism of the emulsifier-free emulsion polymerization of PEO macromonomers (C_1-(EO)$_{24}$-VB, t$C4$-(EO)$_{24}$-VB, C_1-(EO)$_{24}$-MA, tC_4-(EO)$_{24}$-MA, etc.) initiated by AVA was evaluated by Ito et al. [81] (Fig. 3). The higher polymerization rates of the macromonomers were ascribed to the lower rate of diffusion-controlled termination due to their highly crowded segments. The high polymerizability of the amphiphilic PEO macromonomers in water was also attributed to their organization into micelles which locally concentrate and orientate the hydrophobic polymerizing groups for propagation. Besides, the chemically bound surface active groups on the particle surface favor the formation of fine colloidal stable dispersion.

The square-root dependence of [AVA] supports the conventional (instantaneous) bimolecular termination of the growing radicals and/or the higher nucleation activity of the present system compared to the micellar model [9, 10]. The high rate of polymerization resembles the polymerization in particles and was discussed in terms of the micellar polymerization considering that the k_t was extremely low. The linear dependence of R_p on the macromonomer concentration (the first-order dependence of [tC_4-(EO)$_{24}$-MA] or [tC_4-(EO)$_{24}$-VB]) was attributed to the linear dependence of the number of micelles on the macromonomer concentration. This system is transparent, colloidally stable, and macroscopically homogeneous which results from the very small particle size. The very small particles result from the strong association of PEO macromonomers within the micelles or the interface. The macromonomer acts as monomer and emulsifier as well. It takes part in the growth events and in the formation of the polymacromonomer or graft copolymer (or stabilizer).

The rate of polymerization was found to decrease with increasing alkyl chain length from C_1 to C_{18} (hydrophobe) and decreasing EO number (from 60 to 15) in the macromonomers of structure R-$(EO)_n$-VB. The polymerizability of R-$(EO)_n$-VB macromonomers has its maximum (R_p) for n=15–20. Causes of this behavior may be attributed to variations in the end-group spatial dimension and chain end interactions. Thus the macromonomer and its graft copolymer with longer R and shorter PEO chains are highly ordered in micelles or particles. The decrease in the rate of polymerization with increasing PEO chain length was also observed with macromonomers C_{18}-$(EO)_{11-25}$-MA and C_{12}-$(EO)_{14-19}$-MA, respectively [83]. This may be attributed to the decreased radical entry rate and the increased interparticle agglomeration.

The high rates observed in the polymerization of macromonomers H-$(EO)_n$-(α-p-)VB and H-$(EO)_n$-(α-)MA in water initiated by AVA [84] were discussed in terms of a more compact, denser micelle, giving rise to a higher rate of polymerization than the corresponding ω-methoxy derivatives of a comparable degree of polymerization. ω-Methoxy groups were reported to interact with α-ends, causing a somewhat looser organization of amphiphilic monomers. The presence of hydroxyl groups increases the hydrophilicity of PEO chains (or the HLB of the surfactant – the amphiphile) and consequently the stability of polymer particles.

Figure 4 shows that the emulsion polymerization of PEO macromonomers (R-$(EO)_n$-$(CH_2)_m$-St) with hydrophobically enhanced styryl end group is very fast after the start of polymerization [85]. The acceleration of polymerization is more pronounced with a macromonomer containing a larger fraction of hydrophobic units. The strong decrease of the rate with conversion may be attributed to the depletion of free macromonomer micelles or macromonomer. The styryl endgroups characterized with high hydrophobicity are expected to be organized into the more compact micelle cores to enhance the local concentration of unsaturated groups. The polymerization in water with AVA was much faster than that in benzene with AIBN. Since the decomposition rates of AVA in water and AIBN in benzene were found to be only a little different from each other, it was suggested that the compartmentalization of reaction loci is responsible for the fast polymerization in water.

Similarly, the rate of polymerization of PEO-MA macromonomers was found to increase with increasing hydrophobicity around the MA end group in the order [71]

$$C_1\text{-}(EO)_{48}\text{-MA} < C_1\text{-}(EO)_{48}\text{-}C_6\text{-MA} < C_1\text{-}(EO)_{48}\text{-}C_{10}\text{-MA} \tag{35}$$

This result supports the micellar polymerization mechanism, i.e., the stronger hydrophobic intermolecular interactions between PEO molecules favors the

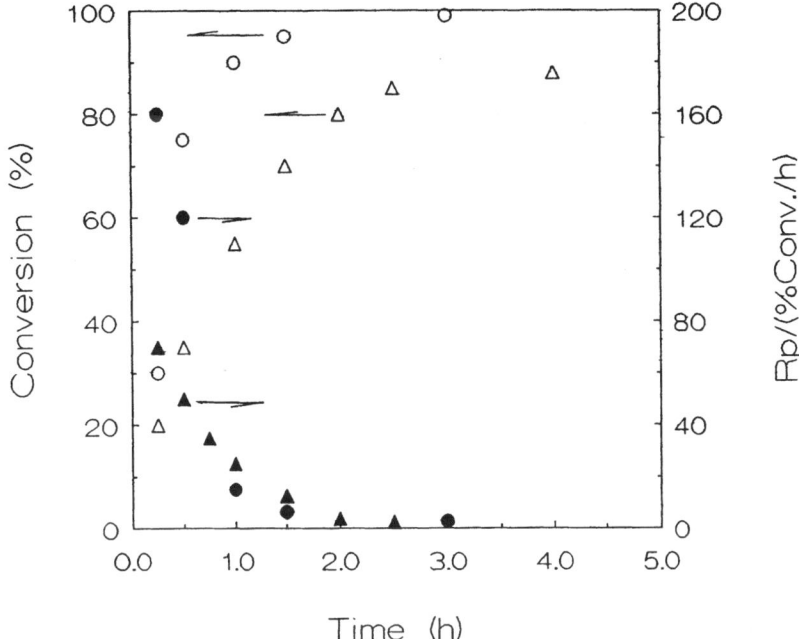

Fig. 4. Dependence of monomer conversion (open symbols) and the rate of polymerization (closed symbols) in the emulsifier-free emulsion polymerization of PEO-VB macromonomers on reaction time and the PEO-VB type [85]. Recipe: [PEO-VB]=0.045 mol dm^{-3}, [AVA]=0.45×10^{-3} mol dm^{-3}, 60 °C. In water: C_1-(EO)$_{38}$-C_7-VB (○,●), C_r-(EO)$_{25}$-VB (△,▲)

formation of larger numbers of micelles (reaction loci). It was found that C_1-(EO)$_{17}$-C_{10}-MA polymerized more rapidly and gave higher final conversions than C_{12}-(EO)$_{19}$-MA. This indicates the importance of the location of the hydrophilic and hydrophobic chains within the PEO macromonomer or relative to the MA group. Thus the location of the hydrophobic alkyl spacer group between PEO and MA favors the association and the growth events. With the C_{12}-(EO)$_{19}$-(MA) macromonomer it is considered that PEO chains take a loop conformation, giving rise to a flower-like micelle, which is entropically unfavored and makes a barrier between the propagating radical and the unsaturated group.

The higher the hydrophilicity of macromonomer, the lower the final conversion. This may be attributed to the formation of hydrophilic or surface active oligomer radicals and the low or high radical entry rate. In the system with hydrophilic C_1-(EO)$_{17}$-MA, the limiting conversion was ca. 60%. Thus the low rate of polymerization at ca. 50 or 60% conversion may be discussed in terms of the solution polymerization, a strong bimolecular termination and the low radical entry rate.

The micellar organization of macromonomers and their polymers or growing radicals in water, together with the high initiation efficiency, enhanced rate of propagation, and reduced rate of termination were supposed to be responsible for the very high rate of emulsion polymerization of PEO-VB (C_1-$(EO)_{48}$-VB) macromonomers [86]. It was found that R_p was more than 50 times higher in water than in benzene. However, the initial rate of polymerization is the maximum one. The square-root dependence on [initiator] was discussed in terms of the conventional bimolecular termination. However, in the terms of the micellar model, the increase in the reaction order above 0.4 is attributed to the delayed termination. The increase in the reaction order above 0.4 may also be discussed in terms of the variation of the surface activity of graft copolymer with its molecular weight (see above). The one-order dependence of R_p on the macromonomer concentration strongly deviates from the micellar model, case 2. This results from the increase of the particle number (reaction loci) with increasing macromonomer concentration. The one-order dependence of R_p on [M] in the classical emulsion polymerization results from the polymerization under the monomer starved conditions and/or in interval 3.

In the aqueous polymerization of C_1-$(EO)_{40}$-C_{11}-MA macromonomer [87], the rate of its polymerization in water was extremely high (within 30 min) as compared to its solution polymerization in benzene (days). It follows first-order kinetics with respect to the macromonomer concentration and one-half-order with respect to peroxodisulfate concentration, respectively. Thus formation of an organized structure, increased local concentration of unsaturated groups in the micelle core, increased number of particles (reaction loci) with macromonomer concentration, and delayed (and depressed) termination are responsible for this behavior. The aqueous system produced polymers with DP_n ranging from about 130 to 160, but it was much lower (DP_n=7) for the benzene system. Thus the increased monomer concentration in particles (micelles) and/or the organized structure macromonomer aggregates favors the growth events. Besides, the activation energy for macromonomer polymerization were 57 kJ mol^{-1} in water but 179 kJ mol^{-1} in benzene. Compartmentalization of reaction loci decreases the termination rate but increases the activation energy of termination. This leads to an increase in both the rate of polymerization and the molecular weight of polymer. However, under the present reaction condition the rate of polymerization is a complex function of the rate of initiation and the particle formation. The number of polymer particles varies with the rate of initiation and the molecular weight of graft copolymers as well. The activation energy (the rate of polymerization) thus varies with the rate of initiation and the particle concentration (the surface activity of graft copolymer is a function of the molecular weight of graft copolymer which decreases with temperature).

Bis-unsaturated PEO macromonomers (VB-$(EO)_n$-VB, n=21-67) having hydrophobic polymerizable vinylbenzyl groups at both ends were found to be very effective in the formation of stable microlatexes [88]. Two different types of the hydrophobic polymerizable VB group were observed in the micelles; within the micelle core or outside the core. The first vinylbenzyl groups were reported to form "strongly organized" micelles. These groups were predominantly consumed during the polymerization and responsible for the high rates of polymerization. Loosely organized unsaturated groups are inactive in polymerization. However, association of PEO macromonomers with n=67 into micelles was depressed. In this case lower rates and molecular weights were observed. As the number of EO units in macromonomer increases, the intermolecular interactions between PEO chains increase. And because of this the solubility of PEO molecules decreases and the viscosity of reaction system (gel formation) increases [89, 90]. This is one reason why the efficiency of PEO macromonomer with very long PEO chains in emulsion polymerization decreases. The particle shell (PEO) barrier for the radical entry is an additional parameter which influences the rate of initiation, the particle nucleation, and growth events. The denser the PEO shell the lower the radical entry rate.

The reaction order with respect to macromonomer concentration was found to be 1.0 for macromonomer with n=21 and 1.2 for macromonomer with n=41, respectively. It increases slightly with increasing number of EO units. This indicates that macromonomer with n=41 is organized in more compact micelles with higher concentration of unsaturated groups and generates more stable polymer particles. For n=67, strong intermolecular interactions (gels) form large aggregates which decreases the amount of monomer for the formation of micelles and growth events.

Molecular weights were very high, increased up to a critical number of EO groups (n=41) and then decreased. In runs with n=21 and 41, the macrogel is also formed. Gelation also took place which is attributed to intermolecular crosslinking.

5.2
Copolymerization of PEO Macromonomers with Styrene

Macromonomers of ω-methoxy PEO undecyl-α-methacrylate (PEO-R-MA) of the structure $CH_3O(-CH_2CH_2O-)_n(CH_2)_{11}-CO_2(CH_3)C=CH_2$ were found to give a stable emulsion with styrene [91]. The relative reactivity $1/r_2$ was found to decrease with increasing PEO chain length (or the number of EO groups (n) in an $(EO)_n$-R-MA macromonomer molecule) and was higher in the emulsion than in the solution. The higher reactivity of macromonomer in water was attributed to the aggregation of macromonomers into micelles and the high local monomer concentration in the polymer particles. The apparent higher reactivity of PEO macromonomer in emulsion copolymerization as compared to solution implies that the polymerization may proceed preferentially at the site between the cores and shells of micelles or particles. It is envisaged that the vinyl groups of macromonomers were concentrated in micellar cores (containing styrene solubilized) which were surrounded by PEO chains extended into aqueous phase to form shells. This situation will effectively remove the repulsion between polystyryl radicals and PEO chains, which retarded the polymerization observed in the solution system.

The average particle diameter (D/nm) for the $(EO)_{40}$-R-MA macromonomer (M_1) increased with increasing wt.% fraction of St in the monomer feed and the particle size distribution varied between 1.14 and 1.01 (D varied between 30 and 90 nm). On the other hand, the latex particles produced by the emulsion copolymerization of $(EO)_{10}$-R-MA or $(EO)_{15}$-R-MA macromonomer ($[M_2]/[M_1]$= 75.2) were much larger and the size decreased with increasing number of EO groups in macromonomer (D varied between 30 and 300 nm).

The M_w values (in 10^{-5} units) increased from 1.1 to 1.4 ($(EO)_{10}$-R-MA), from 1.1 to 1.7 ($(EO)_{15}$-R-MA) and from 2.9 to 4.1 ($(EO)_{40}$-R-MA) with increasing molar ratio [styrene]/[macromonomer] from 39 to 95. Simultaneously, the apparent molecular weight distributions decreased from 3.56 to 2.96, 3.98 to 2.73, and from 1.9 to 1.75, respectively. These data show that the apparent molecular weight of copolymer increases slightly with increasing weight fraction of styrene in the monomer feed.

The ratio M_w/M_n (MWD) decreased with increasing PEO-MA fraction in the monomer feed and/or the number of EO units in the macromonomer. Generally, the M_w/M_n in bulk (homogeneous) systems is a function of the termination mode and the chain transfer events and varies between 1 and 2. In the present disperse systems, MWD is much broader (much above 2) as a result of further contributions, such as polymerization in the continuous phase, interface, and polymer particles. The chain transfer to PEO chains decreased the molecular weight, i,e., the M_w of copolymer decreased with increasing macromonomer concentration and PEO chain length.

A special class of surfactants (macromonomers) derived from maleic anhydride, namely the ethoxyalkyl maleates, with the general structure $CH_3(CH_2)_nOCO-CH=CH-CO(OCH_2CH_2)_mOH$, was very efficient in the preparation of polymer latexes by copolymerizations with styrene (initiated by VA 86 at 70 °C) [92]. The diameter of latex particles increases throughout the reaction and monodisperse latex particles of 200 nm (for m=32) and 300 nm (for m=41) in diameter have been finally produced. High rates of polymerization resulted from the polymerization in micelles. Indeed, the critical micellar concentrations of some macromonomers with dodecyl group as fatty chain and various PEO chain lengths, namely 32 and 41, were found to be 1.79×10^{-5} and 6.22×10^{-5} mol dm^{-3}, respectively, increasing as expected with the PEO chain length.

The emulsifier-free emulsion copolymerization of styrene and poly(methacrylic acid) (PMA) macromonomers

$$R-S-(CH_2-\{C(CH_3)(CO_2H)\}_n-CH_2-(C_6H_5)C=CH_2); \tag{36}$$

2a: $R=C(CH_3)_3$, 2b: $R=HOCH_2CH_2$-

initiated by AVA was fast and the polymer dispersions were stable [93]. The results favor the previous finding that the copolymerization of macromonomer was much faster in water than in solution. Besides, the final conversion reached in homogeneous copolymerization was very small even at 24 h while in the emulsion copolymerization it was ca. 90% conversion (Table 1).

In both systems (2a and 2b) the size of polymer particles decreased and the number of particles strongly increased with increasing macromonomer (neu-

Table 1. Variation of kinetic, molecular weight, and colloidal parameters of dispersion copolymerization of PMA macromonomer and styrene

Run	n	wt%[a]	α_n[b]	R_p[c]	$N \times 10^{-15}$[d]	D_n[e]	D_w/D_n	$M_n \times 10^{-4}$
2a:								
1	6	10	0	1.1				7.5
2	13	10	0	1.8	0.1	1.1	1.21	0.5
2	13	10	0	1.8	0.1	1.1	1.21	0.5
3	13	10	0.2	1.6	14	0.21	1.03	7.4
3	13	10	0.2	1.6	14	0.21	1.03	7.4
4	13	10	0.4	1.7	6.3	0.28	1.07	7.8
4	13	10	0.4	1.7	6.3	0.28	1.07	7.8
5	13	20	0.4	1.2	12.4	0.2	1.03	8.7
6	6	5	0.5	2.4	1.3	0.43	1.04	5.3
7	6	10	0.5	3.5	3.7	0.34	1.02	5.3
2b:								
8	16	10	0	2.6	2.8	0.34	1.06	7.7
9	16	10	0.5	3.6	6.1	0.29	1.03	5.7
10	4	10	0	2	0.3	0.67	1.06	2.6
11	16	10	0	2.6	2.8	0.34	1.06	7.7
2a	16	10	0	1.8	0.1	1.1	1.21	0.5
2b	13	10	0	2.6	2.8	0.34	1.06	7.7

[a] PMA macromonomer
[b] Degree of neutralization of PMA macromonomer
[c] % Conv. h^{-1}
[d] dm^{-3}
[e] µm

tralized or not neutralized) concentration and the decrease in D (or increase in N) was much more pronounced with a neutralized macromonomer. The dependence of the number of particles and particle size on the degree of neutralization in the run 2a was described by a curve with a maximum at 0.2 (10 wt% of macromonomer in the feed). The decrease in the nucleation activity with increasing degree of neutralization ($\alpha_n > 0.2$) is more pronounced in the runs with smaller ratio [macromonomer]/[styrene].

The molecular weights of polymers in the 2a run were found to increase strongly with the degree of neutralization. This may result from the partitioning of monomer and macromonomer between the core and shell of particles. The increase in hydrophilicity of macromonomer decreases the contact between monomer and macromonomer. On the other hand, the increase in hydrophophilicity of macromonomer leads to the compartmentalization of macromonomer. In the run 2b, where the macromonomer is much more hydrophilic, the mutual solubility is low and, therefore, the growth events in particles are suppressed. However, the large α value of macromonomers disfavors the mutual solubility and thus copolymerization and formation of polymers of larger molecular weight.

Poly(ethylene oxide) macromonomers carrying dodecyl, octadecyl, or polystyryl groups at one end and methacryloyl (or p-vinylbenzyl) groups at the other end were used to prepare stable latexes by emulsifier-free emulsion copolymerization with styrene [94]. The maximum rate of emulsion polymerization was observed at low conversions. The high reactivity of the macromonomer in the low conversion range results from the organized association of the macromonomers and graft copolymer molecules (the increase of the local concentration of macromonomer at the reaction loci) and the high monomer concentration. For example, the octadecyl-$(EO)_{35}$-VB macromonomer was found to aggregate by itself to micelle in water. This is the reason why a rapid homopolymerization was observed. At conversions above 50%, the reactivity of macromonomer in the (co)polymer particles decreased and was lower than in solution. This may be ascribed to a low monomer concentration (styrene is almost consumed) and a low radical capture efficiency of polymer particles due to the formation of hydrophilic oligomer radicals in continuous phase. The oligomer radicals are too water soluble to enter the hydrophobic polymer particles.

The values of relative reactivity of macromonomers ($1/r_2$) in both solution and emulsion were found to decrease slightly with conversion. The r_2 values in solution are lower than those for the copolymerization of low-molecular-weight monomers and macromonomers in emulsion. These results support the previous conclusion [95] about the incompatibility of macromonomer with a polymer trunk (polystyrene radical) which suppresses the mutual cross-propagation reactions of comonomers.

The colloidal stability of polymer dispersion prepared by the emulsion copolymerization of R-$(EO)_n$-MA was observed to increase with increasing EO number in the macromonomer [42, 96]. Thus C_{12}-$(EO)_9$-MA did not produce stable polymer latexes, i.e., the coagulum was observed during polymerization. This monomer, however, was efficient in the emulsion copolymerization with BzMA (see below). The C_{12}-$(EO)_{20}$-MA, however, appears to have the most suitable hydrophilic-hydrophobic balance to make stable emulsions. The relative reactivity of macromonomer slightly decreases with increasing EO number in macromonomer. The most hydrophilic macromonomer with ω-methyl terminal, C_1-$(EO)_{39}$-MA, could not disperse the monomer so that the styrene droplets coexisted during polymerization. The maximum rate of polymerization was observed at low conversions and decreased with increasing conversion. The decrease in the rate may be attributed to the decrease of monomer content in the particles (Table 2). In the C_1-$(EO)_{39}$-MA/St system the macromonomer is soluble in water and styrene is located in the monomer droplets. Under such conditions the polymerization in St monomer droplets may contribute to the increase in r_2 values.

Latex particles covered by groups capable of forming hydrogen bonds weakly or moderately lose their association or interaction ability above a specific temperature or a specific electrolyte concentration but recover it reversibly with decreasing temperature. Some EO compounds are typical examples for such groups and their specific properties were studied by Hoshino et al. [97]. PEO-

Table 2. Dependence of the reactivity ratio r_2 and the rate of polymerization in the dispersion copolymerization of PEO-MA macromonomer and styrene (M_2) with the monomer feed composition[a]

Macromonomer type	Conv. (%)	R_p[b]	$[M_2]/[M_1]$	$d[M_2]/d[M_1]$	r_2	$r_{2,sol}$
C_{12}-$(EO)_9$-MA	20	4.0	93	39	0.42	0.79[c]
	15	2.5	48	14	0.3	
	13	2.1	32	5	0.15	
C_{12}-$(EO)_{20}$-MA	5	2.5	329	142	0.43	0.82[c]
	5	2.5	241	110	0.46	
	10	2.5	204	82	0.4	
C_1-$(EO)_{39}$-MA	6	15	501	332	0.66	1.02[d]
	9	30	330	249	0.75	

[a] In water (200 ml), 1 mol% AVA, 2 g monomers, 60 °C
[b] % Conv. h^{-1}
[c] In benzene
[d] In THF

type emulsifiers (PEO-type macromonomers) have their characteristic cloud points dependent on the hydrophile-lipophile balance (HLB) or on the length of PEO chain. Copolymerization of PEO macromonomers with short EO chains (S-PEO) generally resulted in formation of large and uneven shaped particles. On the other hand, PEO macromonomers with the long PEO chains (L-PEO) served to form small, smooth, and spherical copolymer particles. These results were attributed to the different timing of polymerization between hydrophobic and hydrophilic monomers and to the different mode of phase separation of PEO-rich polymer with St-rich polymer in each particle. In the St/PEO-St systems, PEO-rich domains dispersed in the PSt matrix are responsible for the irregular shape of particles. On the contrary, L-PEO-rich polymers, which are more hydrophilic and more flexible, diffuse to the surface of particles during phase separation, forming a surface or a particle shell. The shape of particles was also affected by the monomer feed composition, f_w (Table 3). It was reported that a decrease in f_w made phase separated domains larger and particles less regular.

Electron micrographs of dry particles indicated that the copolymerization of St with L-PEO monomer formed smaller particles but hydrodynamic measurements showed an inverse effect of PEO chain length on the particle size. The increase in hydrodynamic size of L-PEO containing particles cannot be attributed to any aggregation because such particles are more stable than S-PEO containing particles. The ratio of PEO chains in the interior or on the surface of particles and in the feed shows that more hydrophilic L-PEO monomers are apt to remain preferentially in the aqueous medium. Therefore, the apparent surface density of PEO decreased with increasing PEO chain length. A smaller amount of L-PEO monomer can stabilize a larger surface area to form a larger number of particles.

A series of amphiphilic diblock macromonomers were successfully used as steric stabilizers in the emulsion polymerization styrene [98]. Copolymerization led to the formation of polymer latexes of high colloidal stability. These

Table 3. Dependence of the reactivity ratio r_2 and the rate of polymerization in the dispersion copolymerization of C_{12}-(EO)$_9$-MA macromonomer and BzMA (M_2) with the monomer feed composition ($f_w=[M_2]/[M_1]$)[a]

Initiator (mol%)	Conv. (%)	R_p[b]	$[M_2]/[M_1]$	$d[M_2]/d[M_1]$	r_2
KPS	10	20	48.1	65.7	1.37
(0.3)	11	19	19.1	26.8	1.4
	30	45	13.3	18.2	1.37
	8	24	9.1	13.5	1.48
AVA	25	12.5	94.7	12.4	1.31
(1.0)	20	10.0	31.2	44.5	1.43
	15	10.0	18.4	25.3	1.38
	25	12.2	13.5	18.2	1.38
AIBN	9	18.0	92.9	12.4	1.34
(1.0)	20	40.0	32.0	44.5	1.39
AIBN[c]	8	4.0	98.8	40.7	0.41
(1.0)	5	2.5	49.9	19.0	0.38
	5	2.5	31.1	13.1	0.42

[a] In water (200 ml), 2 g monomers, 60 °C
[b] % Conv. h^{-1}
[c] In heptane (40 ml), 4 g monomers, 60 °C

macromonomers are low-molecular-weight poly(ethylene oxide-*b*-propylene oxide) diblock copolymers of variable hydrophobic block sizes. Their M_w varied between 2000 and 6000. The polymerizable functionality, a vinylbenzyl group, is located at the end of the hydrophobic block. Use of these as steric stabilizers for the emulsion polymerization of styrene in conjunction with a water-soluble initiator led to the formation of nonionic colloidal dispersions with good stability towards added electrolyte, even after the latexes had been cleaned with 30% ethanol by serum replacement or centrifugation. The diblocks show increased surface activity (lower surface tensions) as the size of the PPO block increases. The particle diameter decreased as the amount of macromonomer used is increased. The presence of a polymerizable group on the stabilizer greatly improves its covalent grafting to the polymer particles. The surface of polymer particles also contains the adsorbed stabilizer. When the charge-bearing initiators, such as potassium peroxodisulfate, were used, the colloidal stability of the latexes toward added electrolyte was low.

Sutton and Oenick [99] have used PEO-MA macromonomers to prepare graft copolymers and polymer latexes with various functional surface groups. The water insoluble polymers are prepared from oleophilic monomers, monomers bearing groups reactive with free amine or sulfhydryl groups in biological systems ≥0.5 mol% of sulfhydryl groups, and monomers having 0.1–20 mol% of PEO chains. For example, the emulsion terpolymerization of conomomer mixture 85:10:5 styrene:*m*- or *p*-(chloromethyl) styrene:PEO-MA produced the polymer latex having average particle size 0.25 μm.

5.3
Copolymerization of PEO Macromonomers with Other Comonomers

The C_{12}-$(EO)_9$-MA macromonomer was found to be a very effective emulsifier for BzMA in water even at a concentration less than 5 wt%, to give a stable milky emulsion [42, 96]. Table 3 shows that the rate of polymerization depends on the initiator type and polarity of continuous phase. In water solution, the rates are several times higher than in heptane. The rate of polymerization increases with increasing macromonomer concentration in systems with KPS and AIBN, and it is constant with AVA. The higher the macromonomer concentration, the higher the particle concentration and rate of polymerization. These results indicate that distribution of the initiator between the phases influences in complex way the polymerization and nucleation mechanism.

The parameter r_2 is independent of the initiator type for the emulsion, however, and is slightly higher than that obtained in benzene (r_2=1.23) (Table 3). This behavior results from good compatibility of the macromonomer with poly-BzMA. Therefore the reactivity of the macromonomer does not depend so much on the reaction medium type. In contrast, reversed "apparent" reactivity was observed in heptane in which the clear solution of monomer turned into a polymer suspension upon polymerization. Since BzMA is soluble in the medium, it has been suggested that the polymerization occurs preferentially on the (inverse) micelle surface which is enriched by the macromonomers.

The emulsion copolymerization of BA with PEO-MA (M_w=2000) macromonomer was reported to be faster than the copolymerization of BA and MMA, proceeding under the same reaction conditions at 40 °C [100]. Polymerizations were initiated by a redox pair consisting of 1-ascorbic acid and hydrogen peroxide in the presence of a nonionic surfactant (p-nonyl phenol ethoxylate with 20 moles ethylene oxide). In the macromonomer system, the constant-rate interval 2 [9, 10] was long (20–70% conversion). On the other hand, the interval 2 did not appear in the BA/MMA copolymerization and the maximum rate was lower by ca. 8% conversion min^{-1} and it was located at low conversions.

The rate of polymerization was found to be independent of emulsifier concentration around CMC (1.8×10^{-4} mol dm^{-3}) and up to ca. 10^{-3} mol dm^{-3} and then strongly increased with increasing emulsifier concentration (Fig. 5). It can be seen that, for this system, the break in the dependence of the rate on surfactant concentration does not coincide with the CMC of either the surfactant or the surfactant/PEO-MA macromonomer. In fact, these two values are identical at room temperature; at 50 °C the CMC of the surfactant is lower than at 20 °C. The kinetics of particle nucleation for the present nonionic polymerization of BA may not follow a micellar mechanism.

The emulsifier-free emulsion terpolymerization of PEO-MA macromonomer, BA, and acrylic acid (AA) led to the formation of graft copolymers and stable latexes [101]. At the beginning of terpolymerization, the PEO-MA macromonomer polymerized more quickly than BA or AA. Conversion of the macromonomer increased with increasing initiator concentration and with decreasing mo-

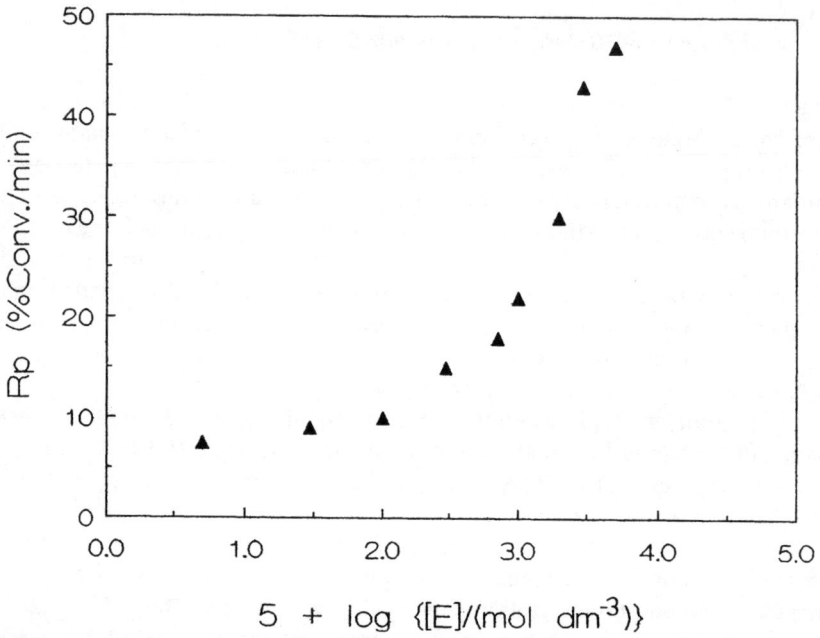

Fig. 5. Dependence of the rate of polymerization [Rp/%Conv./min)] in the free-radical emulsion polymerization of butyl acrylate in the presence of PEO-MA macromonomer on emulsifier (E, p-nonyl phenol ethoxylate) concentration [100]. Temp. 50 °C

lecular weight of the macromonomer or the ratio of macromonomer to monomer in the monomer feed. The IR spectrum of the purified terpolymer showed the presence of H bonds between PEO-MA chains and AA units, while that of the ionomer indicated the disappearance of H bonds and the appearance of carboxylate groups. The average number of macromonomer chains per molecule varied from 6 to 16. The terpolymer could emulsify the C_6H_6 in water and showed high water absorbancy. The terpolymers exhibited crystallinity which diminished with increasing AA content. Tensile strength of the terpolymers increased more noticeably with AA content than with macromonomer content. The mechanical strength of a terpolymer was raised markedly when converted to a metallic salt. The tensile strength of an ionomer containing Na^+ or K^+ was much lower in comparison with an ionomer containing divalent cations because of the stronger complex between PEO-MA and univalent cations.

Ferguson et al. [52] compared the behavior of a range of conventional alkyl ethoxylate surfactants in emulsion polymerizations with their acrylated analogues. This has allowed a direct comparison of identical surfactant structures, one of which remains kinetically mobile in the resultant lattices, while the other becomes chemically bound to the latex particles. The surfactants chosen for this study were C_{12-14}-$(EO)_{30}$ with C_{12-14}-$(EO)_{30}$-A and C_{12-14}-$(EO)_{12}$ with C_{12-14}-

$(EO)_{12}$-A as examples at the extremes of the HLB range available. In all cases both groups of surfactants emulsified the monomers styrene, MMA, and vinyl acetate (VAc). There was no difference between the conventional surfactants and their polymerization analogues with styrene. In the case of the MMA and VAc systems, the polymerizable surfactants yield much less stable emulsions than their non-polymerizable analogues.

The low HLB surfactants C_{12-14}-$(EO)_{12}$ and C_{12-14}-$(EO)_{12}$-A with styrene behaved as with MMA and VAc, although the high HLB species C_{12-14}-$(EO)_{30}$ and C_{12-14}-$(EO)_{30}$-A yielded very stable emulsions. Both conventional and polymerizable alkyl ethoxylates are more effective in stabilizing polystyrene particles than PMMA and PVAc particles. This appears to be related to a greater hydrophobicity of polystyrene, which may simply offer more effective interaction with the hydrophobe of the surfactant. In turn, this may well allow tighter packing at the particle surface and hence increased particle stability. The alkyl ethoxylates with the longer ethoxylate chains (i.e., the higher HLB species) are also more effective. This was discussed in terms of two factors. The first is simply the level of water solubility. The overall level of water solubility imparted to the whole surfactant by the head group must be sufficiently high for stabilization effects to be realized. Clearly, alkyl ethoxylates with longer ethoxylate chains are more water soluble. The second factor concerns the specific mechanism by which alkyl ethoxylates impart stabilization, i.e., the steric stabilization is involved.

It is clear that the polymerizable species are significantly less effective as latex stabilizers than their simple analogues. The polymerizable acrylate group is located at the hydrophilic end of the amphiphile and so its incorporation into macromolecules at the surface of particles would require the hydrophilic EO chain to loop back on itself to allow the hydrophobic alkyl chain also to be adsorbed on the particle surface. Even if this is achievable, the level of stabilization generated from such short hydrophilic loops is likely to be low. The second possible factor explaining the poorer performance of the polymerizable surfactant is the question of the locus of polymerization. The polymerizable acrylic group bound to the hydrophilic PEO chain is located in the particle surface or projects to the aqueous phase. Its participation in growth events may initiate the formation of long polymer chains attached to particle and inter-particle bound association. This may lead to the colloidal destabilization of polymer dispersion. The use of hydrophilic polymerizable PEO monoacrylates (hydrophilic acrylic group bound to PEO chain) as emulsion stabilizers clearly maximizes the chances of the hydrophilic chains being located at the surface of the particles, but also offers the prospect of wastage to polymerization in the aqueous phase.

Larpent and Tandros [102] prepared microlatex particles by polymerization of PEO-MA macromonomer with MMA, styrene, and vinyl acetate. The nonionic latexes are very stable, giving no flocculation up to 6 mol dm^{-3} NaCl or CaCl$_2$ and a critical flocculation concentration (CFC) of 0.6 mol dm^{-3} for Na$_2$SO$_4$ or MgSO$_4$ was estimated. Charged latexes are less stable than the nonionic ones. The CFC of all latexes are determined as a function of electrolyte concentration. With the nonionic latexes, however, the critical flocculation temperature (CFT) is higher than the θ-temperature for PEO at a given electrolyte concentration, indicating enhanced steric stabilization as a result of the dense packing of the chains and hence an elastic contribution to the steric interaction. This is not the case with the charged latex, which shows CFT values lower than the θ-temperature. The charged lattices containing methoxy PEO-MA units are less stable towards the electrolyte.

6
Polymerization of PEO Macromonomers in Other Disperse Systems

The formation of micron-sized polymer particles can be achieved by the suspension polymerization of PEO macromonomers [103–106]. The latexes are sterically stabilized by methoxy poly(ethylene oxide) methacrylate (PEO-MA) moieties, which are covalently grafted on the surface of the particle. For large particles (D=500–900 nm) that have relatively strong van der Waals attractions, the short PEO chains do not provide enough steric stability. In these cases, additional charge must be introduced, for example, by using a charged initiator. The van der Waals attraction was then overcome by a weak steric repulsion from the PEO and a stronger electrostatic repulsion. Thus the polymer particles carried both peroxodisulfate as well as methoxy PEO-MA residual groups.

The macroporous hydrogels were prepared by inverse suspension polymerization (IPS) of PEO-MA (penta and deca EO) macromonomers [107]. The hydrophilic monomers, PEO-MA macromonomer, and N,N-methylenebisacrylamide (MBA, a crosslinking agent) were copolymerized in the presence of stabilizers, KPS, water, and cyclohexane as the oil phase. The authors have tried to find the best conditions for stabilization of water/PEO macromonomer droplets in cyclohexane using nonionic stabilizers having an HLB between 4 and 10. A sorbitan monooleate (Span 80, HLB=4.3) and ABA-block copolymer (ABA), from poly(hydroxystearic acid) and PEO (HLB=7–9) were chosen. The efficiency of the block copolymer appears to be much higher. With Span 80 a good stabilization is only obtained at a relatively high concentration (at 0.1 wt% related to the oil phase) and with ABA already at low concentration (at 0.005 wt%). The bead formation and the dependence of their sizes on concentration of stabilizer follow well the classical behavior [108]. With increasing stabilizer concentration, the bead size decreases and the size distribution becomes narrower. With further increase in the amount of stabilizer, the bead size remains almost unchanged. It was reported that a certain critical concentration of a crosslinker (MBA, ca. 2 mol% with respect to macromonomer) is necessary to get beads. Under such conditions, beads are formed from the beginning of polymerization.

It has been suggested that the mechanism of bead formation occurring in the PEO macromonomer system is quite different from the mechanism proposed by Dimonie et al. [109] for ISP of acrylamide. According to these authors, phase inversion occurs after the start of the reaction. At high conversions, the gel breaks under stirring into small particles which remain as such until the end of polymerization. With PEO macromonomers, beads are present from the beginning up to the end of the reaction and there is no evidence of phase separation. The particular mechanism demonstrated with acrylamide was related to specific solubility properties of the system acrylamide/polyacrylamide/water/cyclohexane.

The micron-sized polymer particles were also prepared by inverse emulsion polymerization and copolymerization of PEO-MA or PEO-A macromonomers with various comonomers (alkyl acrylates and methacrylates) and initiated by both oil- and water-soluble initiators [110]. Various graft copolymers containing ca. 35–95% of comonomers of molecular weigh 1000–500,000 were prepared. The graft copolymers were suggested as stabilizers of polymer dispersion. The efficient graft copolymer (stabilizer) prepared by AIBN initiated solution polymerization containing 30 parts PEO-MA and 70 parts dodecyl acrylate was used in inverse polymerization of sodium acrylate and, acrylic acid in cyclohexane initiated by KPS. Under such conditions, polymer dispersion with very large particle size around 300 μm was prepared.

The preparation of core/shell particles by seed emulsion polymerization was described by Maste et al. [111]. The surface of seed polystyrene particles was modified during polymerization of PEO-MA macromonomer. The polymerization was initiated by KPS at 85 °C. The moment of addition of PEO-MA macromonomer to the polymerization mixture (polystyrene particles) was crucial for the success of copolymerization. These groups are mainly located at the surface when macromonomer is added in the final stage of polymerization. The latexes are composed of relatively large surface charged particles with methoxy PEO groups (containing eight segments of EO, D=480 nm) on the surface. The surface properties of this latex are compared with those of a similar latex but without the PEO surface group. The aggregates formed by coagulation of latex (with PEO) are transient and back-dissociation into single particles occurs. This is not the case with the PEO-free latex. From the complex formation of PEO with molybdatophosphoric acid and proton NMR results, it follows that the average surface coverage of PEO chain is roughly one molecule per square nanometer.

Electrostatic, steric, and the combination of electrostatic and steric mechanisms were investigated by polymerizing a charged styrene saturated polystyrene latex in the absence and the presence of PEO-MA macromonomer [112, 113]. Kinetically, the polymerization under these conditions started at the same rate as in the absence of the macromonomer but proceeded more slowly. The experiments described here indicate that the addition of PEO-MA macromonomer to an emulsion polymerization reaction initiated by sulfate free-radicals leads to the formation of electrosterically stabilized PSt particles. The reaction depends on the time at which PEO-MA is added. Additions at the beginning of the polymerization and in the early stages gave a bimodal particle size distribution whereas additions near the end of the reaction can lead to dispersions with a narrow distribution of particle sizes. For example, the addition of PEO-MA at the beginning of polymerization gave the latex with the dominant population of particles with a diameter of 456 nm, which is considerably larger than the particle size formed in the absence of PEO-MA, i.e., 219 nm. In addition, there was a population of smaller particles present with a broad distribution and a mean diameter of 206 nm. In fact these were not detected when additions were made after 90% conversion.

Ito's group [83] reported the micellar polymerization mechanism was operative during the radical polymerization of PEO macromonomers in cyclohexane and water under similar reaction conditions. The reaction medium has an important effect on the polymerization behavior of macromonomers. Cyclohexane was chosen as a nonpolar type of solvent. The polymerization was found to be independent of the lengths of β-alkyl group (R) and the PEO chain in benzene. On the other hand, the rate of polymerization in cyclohexane increased with increasing number of EO units. This may be attributed to the formation of aggregates (micelles) and/or compartmentalization of reaction loci, i.e., polymerization in distinct aggregates (polymer particles). The C_{12}-$(EO)_{14}$-MA macromonomer polymerized faster in bulk than in benzene but far slower than in water.

The PEO-MA macromonomer is more reactive than the PEO-VB one. Alkyl groups and PEO chain lengths were found to be significant in cyclohexane but the reverse is true in water. The enhanced rate in cyclohexane was ascribed to the inverse micelle formation, since longer PEO chain or alkyl group will favor the aggregation of amphiphilic macromonomer or its polymer (the inverse micelles). The micelle formation in cyclohexane was not detected by laser light scattering measurements. This was ascribed to the formation of somewhat loosely organized inverse micelles which, however, were able to increase the polymerization rate to a rather significant extent. However, the association is supposed to be more pronounced after the start of polymerization and generation of oligomeric or polymeric chains. The agglomeration of amphiphilic polymer chains into polymer particles or micelles accelerate the growth events and, therefore, a higher polymerization rate is observed.

7
Conclusion

The results indicate that amphiphilic PEO macromonomer can act as surfactant, co-surfactant, or monomer. Amphiphilic macromonomers or their graft copolymers present all the typical properties of conventional nonionic surfactants, such as micelle formation, interface tension reduction, and solubilities in both the monomer and polar (water) continuous phase. Water solubility changes according to the hydrophobic content of nonionic emulsifiers. In essence, the HLB value is a direct function of EO content. Higher HLB are therefore indicative of higher water solubility. The PEO macromonomers, under certain conditions, deviate from the current theory of intermolecular interactions of PEO chains. This may be attributed to the intermolecular interactions of PEO chains. Polymeric surfactants assemble into interchain micelles and can also form mixed aggregates incorporating free, monomeric surfactants. At low temperatures, small spherical micelles are formed which aggregate to random clusters with increasing macromonomer concentration and temperature.

The ability of (macro)monomers to get involved in a copolymerization is controlled by the reactivity ratios of the comonomers. The copolymerization parameters, the difference in the molecular weight between the macromonomer and a low-molecular weight comonomer, and the type of reaction media govern the polymerization behavior and the physical properties of graft copolymers given by the distribution of the grafts along the main chain. In most cases the reactivity of macromonomer is inversely proportional to its molecular weight. The polymerization behavior may be attributed to at least three factors – influence of the terminal unsaturated group, influence of the substrate associated with the unsaturated group, and influence of the macromonomer length. The reactivity of the macromonomer functional group mostly follows the reactivity of the low molecular weight monomers carrying the same functional groups. The substrate may promote the agglomeration of surfactant molecules and/or steric factor during chain growth or termination. Different explanations have been put forward to account for the effect of chain length – the reaction growing radical-macromonomer is believed to be diffusion controlled, limitation of the reactive site accessibility is due to the excluded volume effect, and thermodynamic repulsion occurs between growing chains and grafts which are chemically unlike.

When the macromonomer is an amphiphilic polymer, its polymerization in the polar media is unusually rapid as a result of its organization into micelles. Under such conditions, the unsaturated groups are concentrated in the micelle; they mostly form the hydrophobic core of aggregates (micelles). During the polymerization, the non-polymerizing micelles and/or the monomer saturated continuous phase act as a monomer reservoir.

The rate of dispersion (co)polymerization of PEO macromonomers passes through a maximum at a certain conversion. No constant rate interval was observed and it was attributed to the decreasing monomer concentration. At the beginning of polymerization, the abrupt increase in the rate was attributed to a certain compartmentalization of reaction loci, the diffusion controlled termination, gel effect, and pseudo-bulk kinetics. A dispersion copolymerization of PEO macromonomers in polar media is used to prepare monodisperse polymer particles in micron and submicron range as a result of the very short nucleation period, the high nucleation activity of macromonomer or its graft copolymer formed, and the location of surface active group of stabilizer at the particle surface (chemically bound at the particle surface). Under such conditions a small amount of stabilizer promotes the formation of stable and monodisperse polymer particles.

Exponents 0.6–0.8 obtained for the dependence of the rate of dispersion (co)polymerization or molecular weight on initiator concentration were discussed in terms of depressed termination (the first-order radical loss process) and variation of the surface activity of the formed graft copolymer with its molecular weight. The higher the surface activity of graft copolymer (or lower its molecular weight) the higher the particle number.

In the emulsifier free-emulsion polymerization the reaction loci are formed by nucleation of amphiphilic macromomer micelles (micellar mechanism) or by precipitation of growing radicals from the continuous phase (homogeneous nucleation). The micellar mechanism is operative for the emulsion homopolymerization of amphiphilic macromonomers. By taking into account the large number of micelles, there is a high probability that oligomer radicals are captured by micelles before they precipitate from the aqueous phase. In the case of the strongly organized micelles of amphiphilic macromonomers, a high rate of polymerization and large molecular weight polymers are observed. By organization of macromonomers into micelles, unsaturated groups are locally concentrated and oriented. The higher polymerization rates were attributed to the lower rate of termination and/or the compartmentalization of reaction loci. A rapid polymerization does not appear in micelles with loosely organized unsaturated groups of macromonomers. The copolymerization of hydrophilic PEO macromonomers (with low or no association) with hydrophobic low-molecular weight comonomer is controlled by homogeneous nucleation. In this case, the primary particles are formed by precipitation of growing radicals from aqueous phase. The graft copolymer molecules associate with each other to form organized structures (micelle, particles) and, beyond the nucleation period, the oligomeric radicals are efficiently adsorbed by existing premature particles. Reactive emulsifiers formed are chemically bound to the surface of the polymer particles. This strongly reduces the critical amount of surface active groups necessary for the production of a stable particle, desorption of emulsifier from the polymer particles, formation of distinct emulsifier domains during the film formation, and water sensitivity of the latex film.

The square-root dependence of the rate of polymerization on the initiator concentration and the first-order dependence on the macromonomer concentration strongly deviate from the micellar model. The linear dependence of R_p on the macromonomer concentration was attributed to the linear dependence of the number of micelles on the macromonomer concentration. The deviation results from the fact that the macromonomer acts as monomer and emulsifier and/or surface active component is formed during polymerization, i.e., it takes part in the growth and nucleation events. The increase in the reaction order above 0.4 was also discussed in terms of the variation of the surface activity of graft copolymer with its molecular weight.

Acknowledgment. This research is supported by the Slovak Grant Agency (VEGA) (grant number 2/5005/98). The author is also indebted to the Alexander von Humboldt Stiftung for support.

8
References

1. Barrett KEJ (1975) Dispersion polymerization in organic media. Wiley, New York, p 1975
2. Lu YY, El-Aasser MS, Vanderhoff JW (1988) J Polym Sci Part B Polym Phys 26:1187

3. Capek I, Akashi M (1993) JMS Rev Macromol Chem Phys C33:369
4. Tsukahara Y (1994) In: Macromolecular design: concept and practice, macromonomers, chap 5 Characterization, polymerization reactivity and application. Polymers Frontiers International
5. Ito K (1994) In: Macromolecular design: concept and practice, macromonomers, chap 4 Poly(ethylene oxide) macromonomers. Polymers Frontiers International
6. Winzor CL, Mrazek Z, Winnik MA, Croucher MD, Riess MD (1994) Eur Polym J 30:121
7. Paine AJ, Luymes W, Mcnulty J (1990) Macromolecules 23:3104
8. Ober CK, Lok KP (1987) Macromolecules 20:268
9. Hansen FK, Ugelstad J, (1982) In: Piirma I (ed) Emulsion polymerization. Academic Press, New York
10. Fitch RM, Tsai CH (1970) J Polym Sci Polym Chem Ed 8:703
11. Paine AJ (1990) Macromolecules 23:3109
12. Kawaguchi S, Winnik MA, Ito K (1995) Macromolecules 28:1159
13. Lacroix-Desmazes P, Guyot A (1996) Colloid Polym Sci 274:1129
14. Barton J, Capek I (1994) Radical polymerization in the disperse systems. Ellis Horwood, Veda, Bratislava
15. Capek I, Potisk P (1995) Eur Polym J 31:1269
16. Smith WV, Ewart RH (1948) J Chem Phys 16:592
17. Gardon JL (1968) J Polym Sci Part A-1 6:643
18. Gardon JL (1977) Emulsion polymerization, chapter 6. In: Schildknecht CE, Skeist I (eds) Polymerization processes. Wiley, NY
19. Chu H-H, Gau J-H (1995) Makromol Chem Phys 196:2251
20. Hallworth GH, Carless JE (1976) In: Smith AL (ed) Theory and practice of emulsion technology. Academic Press, London, p 305
21. Davis SS, Smith AL (1962) In: Smith AL (ed) Theory and practice of emulsion technology. Academic Press, London, p 325
22. Higuchi WI, Misra J (1962) J Pharm Sci 51:459
23. Delgado J, El-Aasser MS, Silebi CA, Vanderhoff JW (1988) Makromol Chem Macromol Symp 20/21:545
24. Lack CD, El-Aasser MS, Vanderhoff JW, Fowkes FM (1985) In: Shah DO (ed) Macro- and microemulsions, theory and practice. ACS Symp Ser, vol 272, p 272
25. Chamberlain BJ, Napper DH, Gilbert RG (1982) J Chem Soc Faraday Trans 1 78:591
26. Tadros ThF (1984) In: Rosen MJ (ed)Structure/performance relationships in surfactants. 253rd ACS Symposium Series, American Chemical Society, Washington, DC, p 154
27 Guo JS, Vanderhoff JW, El-Aasser MS (1992) J Polym Sci Part A, Polym Chem 30:691
28. Delgado J, El-Aasser MS, Silebi CA, Vanderhoff JW, Guillot J (1988) J Polym Sci Part B, Polym Phys 26:1495
29. Guo JS, Sudol ED, Vanderhoff JW, El-Aasser MS (1992) J Polym Sci Part A, Polym Chem 30:703
30. Guo JS, Sudol ED, Vanderhoff JW, El-Aasser MS (1992) J Polym Sci Polym Chem 30:691
31. Candau F, Leong YS, Fitch RM (1985) J Polym Sci Polym Ed 23:193
32. Rosen MJ (1989) Surfactants and interfacial phenomena, 2nd edn. Wiley, New York
33. Yu GE, Deng YL, Dalton S, Wang QC, Attwood D, Price C, Booth CJ (1992) J Chem Soc Faraday Trans 2537
34. Conway BE, Nicholson JP (1964) Polymer 5:387
35. Liu KJ, Parsons JL (1969) Macromolecules 1:529
36. Tan YY, Challa G (1976) Polymer 17:739
37. Degiorgio V (1985) In: Degiorgio V, Corti M (eds) Physics of amphiphiles: micelles, vesicles and microemulsions. North-Holland: Amsterdam
38. Price CC, Spector R, Tumolo AI (1967) J Polym Sci Part A-1 5:407
39. Bannister DJ, Davies GR, Ward IM, McIntyre JE (1984) Polymer 25:1600
40. Trijasson P, Frere Y, Gramain P (1990) Makromol Chem Rapid Commun 11:239

41. Frere Y, Gramain Ph (1992) Reactive Polym 16:137
42. Ito K, Yokoyama S, Arakawa F (1986) Polym Bull 16:345
43. Hashimoto K, Sumimoto H, Kawasumi M (1985) Polym J 17:1045
44. Loffler R, Richtering WR, Finkelmann H, Burchard W (1992) J Phys Chem 96:3883
45. Richtering WH, Burchard W, Jahns E, Finkelmann H (1988) J Phys Chem 92:6032
46. Richtering W, Loffler R, Burchard W (1992) Macromolecules 25:3642
47. Imae T, Sasaki M, Ikeda S (1989) J Colloid Interface Sci 127:511
48. Miyagishi S, Kurimoto H, Asakawa T (1995) Langmuir 11:2951
49. Kawaguchi S, Yekta A, Duhamel J, Winnik MA (1994) J Phys Chem 98:7891
50. Winnik FM, Winnik MA, Ringsdorf H, Venzmer J (1991) J Phys Chem 95:2583
51. Ito K, Tanaka K, Tanaka H, Imai G, Kawaguchi S, Itsuno S (1991) Macromolecules 24:2348
52. Ferguson P, Sherrington DC, Gough A (1993) Polymer 34:3281
53. Cohin D, Zana R, Candau F (1993) Macromolecules 26:2765
54. Franks F (1975) In: Frank F (ed) Water: a comprehensive treatise, vol 4. Plenum, New York
55. Tiddy GJD, Walsh MF (1983) Stud Phys Theor Chem 26:151
56. Tuzar Z, Kratochvil P (1976) Adv Colloid Interface Sci 6:201
57. Riess G, Duplessix R, Gallot Y, Pichot C (1979) Macromolecules 12:1180
58. Barker MC, Vincent B (1984) Colloids Surf 4:297
59. Franta E (1966) J Chim Phys 63:595
60. Eicke HF, Christen H (1978) Helv Chim Acta 61:2258
61. Cogan KA, Gast AP (1990) Macromolecules 23:745
62. Riess G, Rogez D (1982) Polym Prepr (Am Chem Soc Div Polym Chem) 23:19
63. Hruska Z, Winnik MA, Hurtrez G, Riess G (1990) Polym Commun 31:402
64. Wilhelm M, Zhao CI, Wang Y, Xu R, Winnik MA (1991) Macromolecules 24:1033
65. Capek I, Riza M, Akashi M (1992) Polym J 24:959
66. Capek I, Riza M, Akashi M (1995) Eur Polym J 31:895
67. Capek I, Murgasova R, Berek D (1997) Polym International 44:174
68. Yekta A, Duhamel J, Adiwidjaja H, Brochard P, Winnik MA (1993) Langmuir 9:881
69. Capek I, Riza M, Akashi M (1992) Makromol Chem 193:2843
70. Capek I, Riza M, Akashi M (1997) J Polym Sci (in press)
71. Furuhashi H, Kawaguchi S, Itsuno S, Ito K (1977) Colloid Polym Sci 275:227
72. Riza M, Capek I, Kishida A, Akashi M (1993) Angew Makromol Chem 206:69
73. Brandrup J, Immergut EH (eds) (1989) Polymer handbook, 3rd edn. Wiley, New York
74. Lacroix-Desmazes P, Guyot A (1996) Macromolecules 29:4508
75. Bromley CWA (1986) Collids Surf 17:1
76. Lacroix-Desmazes P, Guyot A (1996) Polym Bull 37:183
77. Lacroix-Desmazes P, Guyot A (1996) Colloid Polym Sci 274:1129
78. Nugroho MB, Kawaguchi S, Ito K, Winnik MA (1995) J Macromol Sci Pure and Appl Chem (Macromol Reports) A32:593
79. Inoue T, Kimi T, Inagaki K, Suzuki S Jpn Kokai Tokkyo Koho JP 03 95,2209
80. Capek I (1995) J Macromol Sci Pure and Appl Chem (Macromol Reports) A32:749
81. Ito K, Tanaka K, Imai G, Kawaguchi S, Itsuno S (1991) Macromolecules 24:2348
82. Rempp PF, Franta E (1994) Adv Polym Sci 58:1
83. Ito K, Kobayashi H (1992) Polym J 24:199
84. Ito K, Hashimura K, Itsuno S, Yamada E (1991) Macromolecules 24:3977
85. Chao D, Itsuno S, Ito K (1991) Polym J 23:1045
86. Nomura E, Ito K, Kajiwara A, Kamachi M (1997) Macromolecules 30:2811
87. Liu J, Chew CH, Gan LM (1996) JMS Pure Appl Chem A33:337
88. Shibata Y, Kawaguchi S, Ito K (1992) Polym Prepr Japan 41, 28 D-09
89. Tadokoro H (1967) Macromol. Rev. 1:119
90. Zgoda MM (1991) Acta Polon Pharm-Drug Res 48:67
91. Liu J, Chew CH, Wong SY, Gan LM (1996) JMS Pure Appl Chem A33:1181

92. Hamaide T, Zicmanis A, Monnet C, Guyot A (1994) Polymer Bull 33:133
93. Ito K, Sabao K, Kawaguchi S (1995) Macromol Symp 91:65
94. Ito K, Arakawa F, Tanaka H, Hashimura K, Itsuno S (1989) Rep Asahi Glass Found Ind Technol 54:105
95. Ito K, Tsuchida H, Hayashi A, Kitano T, Yamada E, Matsumoto T (1985) Polym J 17:827
96. Ito K, Tsuchida H, Kitano T (1986) Polym Bull 15:425
97. Hoshino F, Sakai M, Kawaguchi H, Outsuka Y (1987) Polym J 19:383
98. Schechtman LA (1992) Stud Polym Sci 11:23
99. Sutton RC, Oenick MDB Eur Pat Appl EP 468,584
100. Westby MJ (1988) Colloid Polymer Sci 266:46
101. Xie H, Liu X, Guo J (1990) Eur Polym J 26:1195
102. Larpent C, Tandros TF (1991) Colloid Polym Sci 269:1171
103. Cowell C, Lin-in-on R, Vincent B (1978) J Chem Soc Faraday Trans 74:337
104. Bromley CWA (1986) Colloids Surfaces 17:1
105. Liang W, Tadros ThF, Luckham PF (1992) J Colloid Interface Sci 153:131
106. Goodwin JW, Hearn J, Ho CC, Ottewill RH (1974) Colloid Polym Sci 252:464
107. Trijasson P, Frere Y, Gramain P (1990) Makromol Chem Rapid Commun 11:239
108. Dawkins JV (1989) In: Allen G (ed) Comprehensive polymer science, vol 4. Pergamon, New York, p 231
109. Dimonie MV, Boghina CM, Marinescu NN, Marinescu MM, Cincu CL, Oprescu CG (1992) Eur.Polym J 18: 639
110. Nagasuna K, Nanba T, Kimura K, Shimomura T, Rakuya K, Hozumi Y Jpn. Kokai Tokkyo Koho JP 02,115,201
111. Maste MCL, van Velthoven APCM, Norde W, Lyklema J (1994) Colloids Surfaces A: Physicochem Eng Aspects 83:255
112. Ottewill RH, Satgurunthan R (1987) Colloid Polym Sci 265:845
113. Ottewill RH, Satgurunthan R (1995) Colloid Polym Sci 273:379

Editor: Prof. K. Dušek
Received: April 1998

Molecular Engineering of π-Conjugated Polymers

Jerry L. Reddinger, John R. Reynolds*

Department of Chemistry, Center for Macromolecular Science and Engineering, George and Josephine Butler Polymer Research Laboratories, University of Florida, Gainesville, FL 32611, USA
*e-mail: reynolds@chem.ufl.edu

An extensive review of the synthesis of π-conjugated polymers is presented using a tutorial approach to provide an introduction to the field intended for the undergraduate student and the experienced chemist alike. The many synthetic methodologies that have been used for the synthesis of conjugated polymers are outlined for each class of polymers with a focus on research from the 1990s. The effect of structure on electrical properties is detailed. Specific systems reviewed include the polyacetylenes, polyanilines, polypyrroles, polythiophenes, poly(arylene vinylenes), and polyphenylenes.

Keywords: Conducting polymers, Conjugated polymers, Polyacetylenes, Polyanilines, Polypyrroles, Polythiophenes, Poly(arylene vinylenes), Polyphenylenes

1	Introduction	59
1.1	Our Mission	59
1.2	Why Conjugated Polymers?	59
1.3	Applications of π-Conjugated Polymers	60
1.4	Focusing on Synthesis	61
1.5	Other Literature	61
2	**The Conducting Polymer**	62
2.1	The Extended π-System	62
2.2	Doping and Electrical Conductivity	65
2.3	Optical Properties	67
3	**Polymerization Methods**	67
3.1	Introduction	67
3.2	Electropolymerization	68
3.3	Lewis Acid-Induced Polymerization	68
3.4	Ring-Opening Metathesis Polymerization	70
3.5	Transition Metal-Catalyzed Coupling Polymerizations	70
4	**Polyacetylene**	72
4.1	Direct Polymerization of Acetylene	72
4.2	Substituted Polyacetylenes	73
4.3	Cyclic Polyacetylene Derivatives	73

4.4	Water-Soluble Polyacetylenes	74
4.5	Precursor Polyacetylene Via Ring Opening Metathesis Polymerization (ROMP)	75
5	**Poly(*p*-phenylene)**	**77**
5.1	Direct Synthesis of Poly(*p*-phenylenes)	78
5.2	Soluble Substituted PPPs	78
5.3	Soluble, Fully Aromatic Polyphenylenes	80
5.4	Polyphenylenes Via Soluble Precursor Polymers	82
5.5	Water-Soluble Poly(*p*-phenylenes)	83
5.6	"Planarized" Polyphenylenes	84
6	**Poly(phenylene vinylene)**	**86**
6.1	Introduction	86
6.2	PPV Via Classical Organic Means	86
6.3	Direct Synthesis of PPV Via ADMET	88
6.4	Precursor Polymer Route	89
6.5	The Heck Reaction	93
7	**Polythiophenes**	**96**
7.1	Introduction	96
7.2	Lewis Acid-Induced Polymerizations	97
7.3	Transition Metal-Mediated Coupling Polymerization	98
7.4	Poly(alkylthiophenes)	99
7.5	Regioregular Polythiophenes	101
7.6	Poly(3,4-ethylenedioxythiophene) (PEDOT) and Derivatives	103
8	**Polypyrroles**	**104**
9	**Polyaniline**	**106**
10	**New Vistas**	**109**
10.1	Copolymers	109
10.2	Conjugated Oligomers	110
10.3	Cyclization of Prepolymers	110
10.4	Silole Systems	111
10.5	Transition Metal-Containing CPs	111
11	**Perspectives**	**113**
12	**References**	**113**

1
Introduction

1.1
Our Mission

The past two decades have produced a number of new additions to the chemist's vernacular, and the term "conducting polymer" has proven to be one of the most visible. Literally thousands (tens of thousands?) of articles have been published concerning the synthesis and properties of these novel materials, along with many stemming from an applications arena. But what are these plastics that possess the ability to carry electrical conductivity, and how are they synthesized? This article is written with the aim of answering this question. In concept, we have utilized the tutorial approach hoping to provide an introduction to the field intended for the undergraduate student and the experieced chemist alike. Because of the large amount of work dedicated to this aspect of the field, many conjugated polymers are commonly referred to as "conducting polymers". While the main focus of this review concentrates on synthetic avenues leading to π-conjugated polymers (CPs), the presentation of some of the "nuts and bolts" (while dispensing with the esoteric jargon) will help bring even the uninitiated novice up to speed. So, on this note, we begin.

1.2
Why Conjugated Polymers?

To put it mildly, a significant research effort has been dedicated recently to the synthesis of fully conjugated organic polymers [1]. Custom-tailored synthetic methodologies have been developed for the preparation of these historically difficult to synthesize materials. Unique synthetic conditions have been employed to obtain materials with high molecular weight, specifically controlled macromolecular architectures, high levels of purity, and useful processability. The motivation for this research effort lies in the novel properties and potential applications demonstrated by these polymers. The neutral polymers can behave as semi-conductors, exhibiting optoelectronic properties analogous to their typical inorganic counterparts (e.g. silicon, GaAs, etc.). These properties are easily varied and controlled by the main chain polymer structure. Redox chemistry, commonly called oxidative or reductive doping, allows conversion of the neutral conjugated polymers into their charged, or doped, forms. There is a concomitant increase in the electronic conductivity; in some cases reaching values comparable to true metals. In addition, this redox doping process is reversible, allowing the polymers to be repeatedly switched between their neutral and charged states. This leads to the modification of many properties including conductivity, electromagnetic absorption, luminescence, paramagnetism, ion content and volume.

1.3
Applications of π-Conjugated Polymers

Numerous applications have been demonstrated and proposed for conjugated polymers. Some of the present and potential commercial applications of these systems are listed below [2].

- Storage batteries/supercapacitors/electrolytic capacitors/fuel cells
- Sensors
  - Biosensors
  - Chemical Sensors
- Ion-specific membranes
- Ion supply/exchange devices
  - Drug and biomolecule release
- Electrochromic displays
  - Electromagnetic shutters
- Corrosion protection
- Transparent conductors
- Mechanical actuators
  - Artificial muscles
- Gas separation membranes
- Conductive Thermoplastics
- Microwave weldable plastics
- EMI shielding
- Aerospace applications
  - Lightning strike protection
  - Microwave absorption/transmission
- Conductive textiles
- Anti-static films and fibers
  - Photocopy machines
- Conductor/insulator shields
- Neutron detection
- Photoconductive switching
- Conductive adhesives and inks
- Electronics
  - Conductor feedthroughs
- Non-linear optics
- Electroluminescence
- Electronic devices

This list can be divided into three main classes based mainly on function and redox state. First, applications that utilize the conjugated polymer in its neutral state are often based around their semi-conducting properties, as in electronic devices such as field effect transistors or as the active materials in electroluminescent devices. Secondly, the conducting forms of the polymers can be used for electron transport, electrostatic charge dissipation, and as EMI-shielding materials. These first two types of applications can be viewed as "static" applications (as the polymers do not change their electronic state during use). The final area of applications is based around those that use the ability of the polymers to redox switch between charged states. These include their use as battery electrode materials, electrochromic materials, and in ion release devices and biosensors.

1.4
Focusing on Synthesis

Convenient and reproducible syntheses are required before materials can be developed for a particular application. In recent times an extensive worldwide research effort has been specifically directed towards synthetic aspects and is the focus of this chapter. It is the goal of this work to bring together, in one body, the many synthetic methodologies that have been used for the synthesis of conjugated polymers. It is well known that synthetic polymers are structurally complicated in aspects of molecular weight, polydispersity, branching, cross-linking, and structural defects along the main chain. Important properties can be highly affected by the presence of various structural irregularities, making the intricacies of the synthetic preparation employed crucial to understanding the fundamental physics of the system and their usefulness for any application. In addition, comparisons of a polymer between laboratories are often made with little understanding of the full ramifications of structure on the resulting properties.

This chapter is divided into ten sections with distinctions based on the parent for each family of polymers. Within each section many derivatives are presented with each example representing a specific and unique polymer or material in its own right. We hope this organization allows the reader to quickly determine the synthetic method of choice.

1.5
Other Literature

Numerous reviews have been published on various aspects of the synthesis, properties and applications of conjugated polymers [1]. For these reasons we have chosen to focus our review of synthetic developments on research from the 1990s (although many older examples will be included to give a more complete picture). The earlier reviews nicely address previous developments of the field, and we hope that this chapter will bring the reader up to the state of the art in the field. Due to the large amount of work published in the field, it is impossible to reference each and every article. The availability of referenced material to the general reader was considered during the preparation of this manuscript. Accordingly, we have chosen not to cite material from the patent literature, as we are well aware of the frustration caused by elusive key references. In a field such as this, where rapid development has continued for a number of years, it is difficult to ensure that each contribution is noted. We apologize in advance for any omissions.

2
The Conducting Polymer

2.1
The Extended π-System

In the late 1970s Heeger and MacDiarmid found that polyacetylene [$(CH)_n$] produced by Shirikawa's method exhibited a 12 order of magnitude increase in electrical conductivity when exposed to oxidizing agents. Since that discovery, a vast array of other CPs have been synthesized. The most common of these, in addition to polyacetylene, are shown below in Scheme 1.

Apart from polyaniline (which will be discussed in detail later), all of these systems share one common structural feature, namely a rigid nature brought about by the sp^2 carbon-based backbone. The utilization of the conjugated construction affords polymer chains possessing extended π-systems, and it is this feature alone that separates CPs from their other polymeric counterparts. Using this generic, lowest energy (fully bonding) molecular orbital (MO) representation as shown by the π-system model, the picture of primary concern that is generated by these networks consists of a number of π and π* levels (Fig. 1). However, unlike the discreet orbitals that are associated with conjugated organic

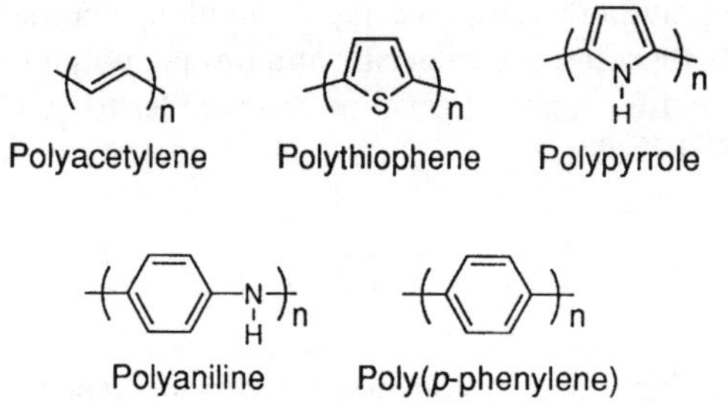

Scheme 1

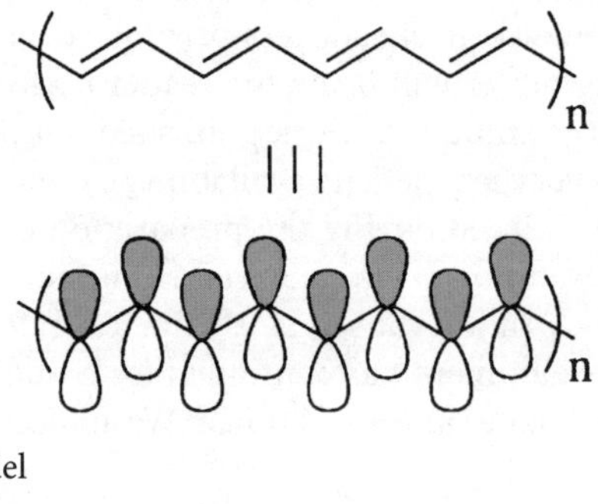

Fig. 1. The π-system model

molecules, the energy of the polymers' MOs are so close in energy that they are indistinguishable. In fact, for long chains orbital separation is so small that band formation occurs as illustrated by the MO diagram (Fig. 2).

The electrical properties of any material are a result of the material's electronic structure. The presumption that CPs form bands through extensive molecular obital overlap leads to the assumption that their electronic properties can be explained by band theory. With such an approach, the bands and their electronic population are the chief determinants of whether or not a material is conductive. Here, materials are classified as one of three types shown in Scheme 2, being metals, semiconductors, or insulators. Metals are materials that possess partially-filled bands, and this characteristic is the key factor leading to the conductive nature of this class of materials. Semiconductors, on the other hand, have filled (valence bands) and unfilled (conduction bands) bands that are separated by a range of forbidden energies (known as the "band gap"). The conduction band can be populated, at the expense of the valence band, by exciting electrons (thermally and/or photochemically) across this band gap. Insulators possess a band structure similar to semiconductors except here the band gap is much larger and inaccessible under the environmental conditions employed.

At first glance one might necessarily expect that the π electrons of polyacetylene would produce a half-filled band and result in the polymer being metallic in nature. However, the one-dimensional nature of the polyacetylene chain leaves it susceptible to an instability that forces the polymer to retain its strict, alternating series of long and short bonds. This instability, analogous to a "Peierl's distortion", is very common among molecular solids and is the result of the coupling of electrons with phonons (lattice vibrations) [3]. Given the relatively soft nature of the lattice in such low dimensional solids, the total energy of the system can be decreased through a doubling of the unit cell, concomitantly opening a gap in the conduction band at the Brillouin zone boundary. In fact, structural studies of

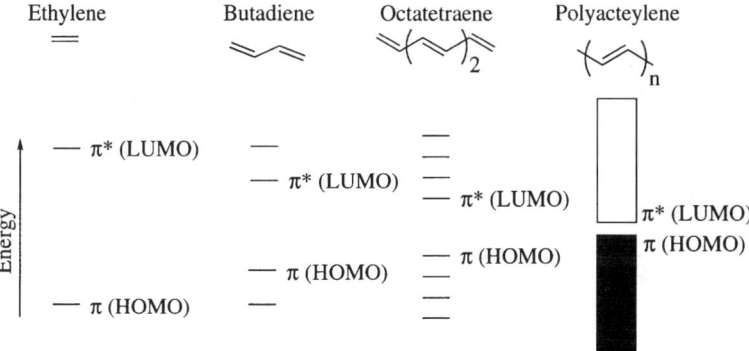

Fig. 2. Molecular orbital (MO) diagram

polyacetylene have shown the polymer to possess a localized backbone consisting of alternating long and short bonds [4]. This is in stark contrast to aromatic molecules, such as benzene, where the bonds are completely delocalized.

It is the Peierl's instability that is believed to be responsible for the fact that most CPs in their neutral state are insulators or, at best, weak semiconductors. Hence, there is enough of an energy separation between the conduction and valence bands that thermal energy alone is insufficient to excite electrons across the band gap. To explain the conductive properties of these polymers, several concepts from band theory and solid state physics have been adopted. For electrical conductivity to occur, an electron must have a vacant place (a hole) to move to and occupy. When bands are completely filled or empty, conduction can not occur. Metals are highly conductive because they possess unfilled bands. Semiconductors possess an energy gap small enough that thermal excitation of electrons from the valence to the conduction bands is sufficient for conductivity; however, the band gap in insulators is too large for thermal excitation of an electron accross the band gap.

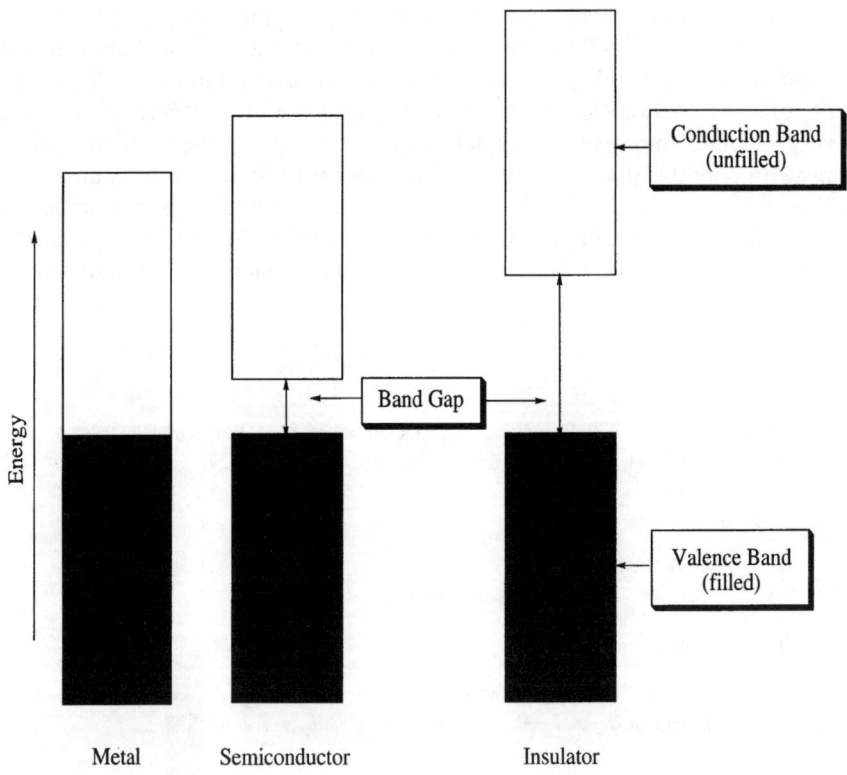

Scheme 2

2.2
Doping and Electrical Conductivity

So how does this 12 order of magnitude increase in electrical conductivity for polyacetylene occur? The diffuse nature of the extended π-system readily allows electron removal from, or injection, into the polymer. The term "doping" has been borrowed from semiconductor physics with "p-doping" and "n-doping" being used to describe polymer oxidation and reduction, respectively. Doping in regards to semiconductors is quite different as it is a very distinct process carried out at low levels (<1%) as compared to CP doping (usually 20–40%). However, the manner by which doping transforms a neutral CP into a conductor remained a mystery for many years.

EPR studies have shown that both the neutral and heavily doped CPs possess no net spin, interpreted as no unpaired electrons, while moderately doped materials were discovered to be paramagnetic in nature. Conductivity experiments showed that it was the "spin-less", heavily-doped form that is the most conductive for a given CP. Such behavior marks an abrupt departure from simple band theory, which centers around spin-containing charge carriers.

Polyacteylene turns out to be a special case when considering its neutral and doped forms. Comparison of the two neutral forms, shown in Scheme 3, reveals them to be structurally identical, and thus, their ground states are degenerate in energy. Two successive oxidations on one chain could yield radical cations that, upon radical coupling, become non-associated charges termed positive "solitons".

In contrast to polyacetylene, the other CPs shown in Scheme 1 have non-degenerate ground states (i.e. they do not possess two equivalent resonance forms), and thus, do not show evidence of soliton formation. In this instance, the oxidation of the CP is believed to result in the destabilization (raising of the energy) of the orbital from which the electron is removed. This orbital's energy is

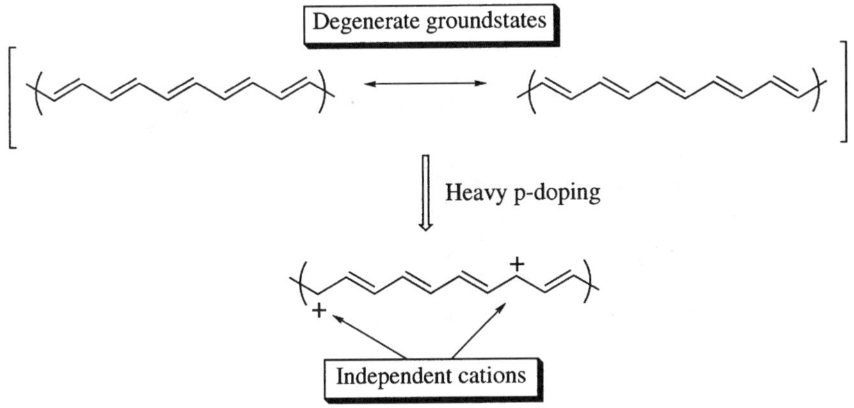

Scheme 3

increased and can be found in the energy region of the band gap as shown in Scheme 4. Initially, if only one electron per level is removed a radical cation is formed and is known as a "polaron" (Scheme 4b). Further oxidation removes this unpaired electron yielding a dicationic species termed a "bipolaron" (Scheme 4c). High dopant concentrations create a bipolaron-"rich" material and eventually leads to band formation of bipolaron levels. Such a theoretical treatment, thereby, explains the appearance, and subsequent disappearance, of the EPR signal of a CP with increased doping as the neutral polymer transitions to the polaronic form and subsequently to the spinless bipolaronic state.

Contrary to polyacetylene's independent charges, the bipolaron unit remains intact and the entire entity propagates along the polymer chain. Scheme 5 shows this behavior using polythiophene as an example. In the case of unsubstituted polythiophene, the bipolaronic unit is believed to be spread over six to eight rings. This "bipolaron length" is by no means an absolute number as different polymer backbone and substituent types yield various lengths.

While this general model for charge carrier generation has developed over the years, it is not without conjecture. As one alternate possibility, the presence of diamagnetic π-dimers, resulting from the combination of cation radicals, has

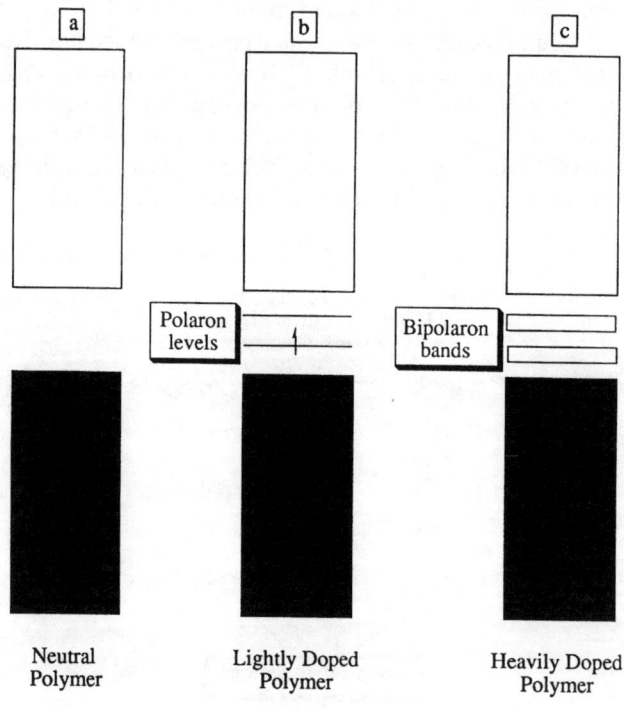

Scheme 4

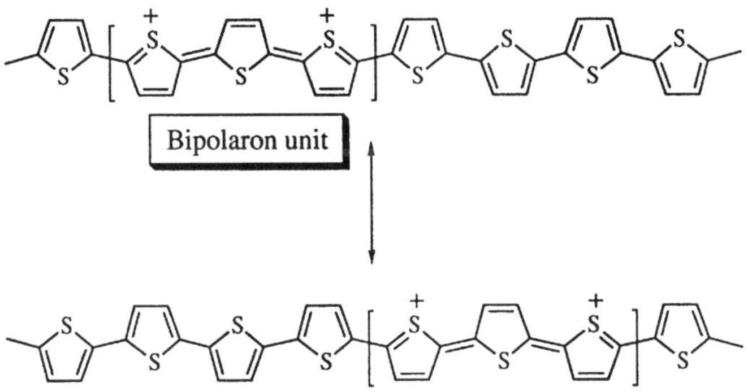

Scheme 5

been proposed [5–7]. Much of the basis for these theories comes from investigations into the structural and electronic properties of small conjugated molecules.

2.3
Optical Properties

Doping also brings about radical changes in a CP's optical properties. For instance, neutral polythiophene films are red in color, while doped polythiophene is blue in color. A broad variety of color changes that can be structurally controlled have been observed for the CPs in changing between their respective redox states. These optical changes are a consequence of polaronic levels and bipolaron bands residing in the band gap. While the neutral polymer only has its charcteristic π-π^* transition, several new transitions are possible to the orbitals in the bipolaronic state. The energies of these new transitions are necessarily lower and result in the polymer having red-shifted absorptions. While the altering of a CP's optical properties can be readily accomplished via chemical means, electrochemical doping is attractive from an applications standpoint, and these polymers provide a new family of electrochromic materials.

3
Polymerization Methods

3.1
Introduction

While there are many polymerization reactions utilized to produce a variety of π-conjugated polymers, four general C-C bond-forming schemes have afforded many of the polymers found in the literature and continue to be valuable routes. These four methodologies, namely electrochemical synthesis, Lewis acid-induced polymerization, ring-opening metathesis and transition metal-catalyzed coupling polymerizations, will be discussed in this section with the aim of presenting their broad utility.

3.2
Electropolymerization

Electrochemical synthesis utilizes the ability of a monomer to be self-coupled upon irreversible oxidation (anodic polymerization) or reduction (cathodic polymerization). While this method does not always produce materials with well-defined structures (as do the three other polymerization methods to be discussed), electropolymerization, nonetheless, is a rather convenient alternative, avoiding the need for polymer isolation and purification. Of these two routes, anodic polymerization is the most widely explored as monomers such as pyrrole and thiophene are relatively electron-rich and prone to oxidation. For this reason the anodic route will be the focus of the remainder of this presentation.

While it is well accepted that anodic electrosynthesis of conducting polymers begins with monomer oxidation to form an ion-radical reactive intermediate, there remains significant conjecture as to the ensuing steps leading to oligomer/polymer formation. As shown in Scheme 6, chain growth of the oxidized species can occur via cation radical coupling (path A) or through electrophilic addition of the the oxidized monomer by an unaffected moiety.

3.3
Lewis Acid-Induced Polymerization

Lewis acid-induced polymerization is also a relatively convenient means to obtain conducting polymers. As with anodic electropolymerizations, monomers are oxidized by an external source, but in this instance electron transfer occurs from the monomer or growing chain to the chemical oxidant. Of the various Lewis acids that are readily available, $FeCl_3$ has become the most popular choice given its effectiveness and low cost. Scheme 7 shows a typical polymerization where the resulting polymer is obtained in its conducting state with $FeCl_4^-$ dopant ions.

This form of chemical polymerization can also produce materials possessing coupling defects that reduce performance owing to the lack of selectivity of the oxidant. (This problem will be discussed further in following sections.) Unlike electrochemical synthesis, though, polymers prepared via chemical oxidants must be isolated and dedoped, which is typically effected using reducing agents such as NH_3 or N_2H_2. The benefit of the Lewis acid-induced polymerization is that it can be used to produce large quantities of material with few difficulties encountered by scale-up. Furthermore, with proper functionalization, soluble conjugated polymers can be obtained that are more amenable for many applications than are the corresponding infusible electrosynthesized films, or insoluble powders, of non-functionalized polymers.

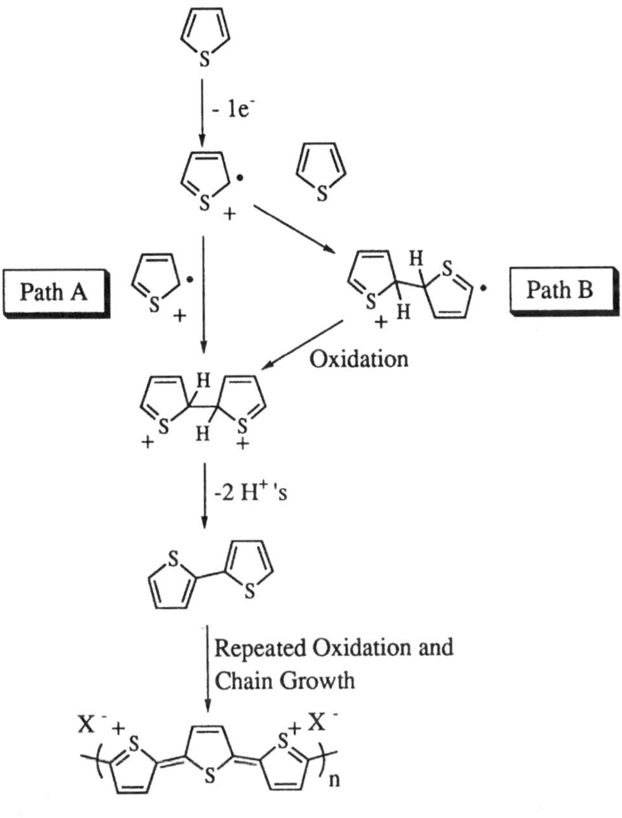

Scheme 6

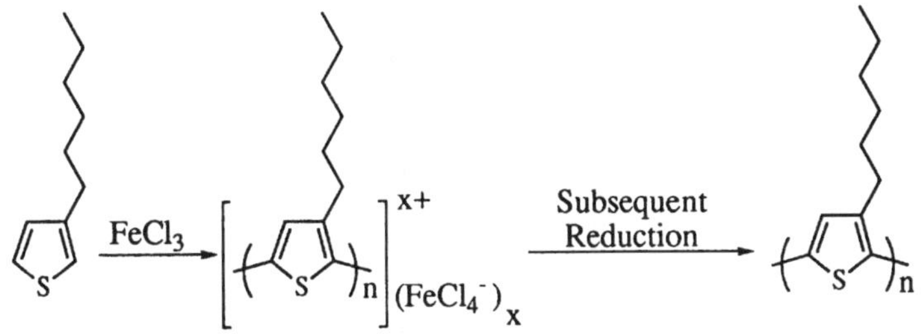

Scheme 7

3.4
Ring-Opening Metathesis Polymerization

Olefin metathesis is a staple of organometallic chemistry and introductions to the subject can be found in almost all introductory organometallic textbooks [8-11]. The premise of metathesis operates on the ability of certain reactive organometallic species (usually carbene or alkylidene complexes) to add to double bonds forming metallacycles. These cyclic species are in constant equilibrium with unreacted olefin and starting complex, and thus, toggle back and forth between the metallacyclic and olefin forms. However, the olefin that is obtained by reversal of the forward reaction need not be the original one. As the goal of olefin metathesis is the formation of "new" unsaturated molecules, the term productive metathesis is used to describe the formation of desired products while degenerate metathesis yields the original olefin and carbene complex. The tendency of a reaction to undergo productive metathesis can be enhanced if one of the olefins produced can be removed from the reaction mixture upon its formation. Thus, the entropic advantages of the elimination of ethylene or propylene by the use of high vacuum can be exploited to afford new olefin in exceptionally high yields.

To increase the efficiency of the reaction, and the molecular weight of the materials produced, cyclic olefins, possessing a large degree of ring strain (an example is given in Scheme 8), are often employed. With such a "ring-opening metathesis polymerization" (ROMP) strategy, the enthalpic change accompanying ring opening drives the reaction to yield polymers with very high molecular weights. Further advances in ROMP have produced systems that avoid the production of volatile by-products, yet still afford materials with the desired properties [8, 10].

3.5
Transition Metal-Catalyzed Coupling Polymerizations

The final route to conducting polymers to be described also involves organometallic processes. Of these coupling reactions, the cross-coupling polymerization is, perhaps, the best explored and most widely employed. Two monomer func-

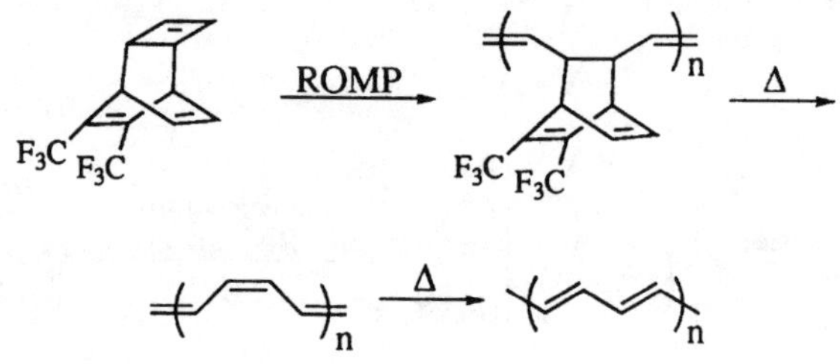

Scheme 8

tionalities, each playing key roles in the catalytic cycle, are required in such schemes, and they are commonly halogens (or halide equivalents including tosylates and triflates) and electropositive metal-containing groups (-B(OR)$_2$, -MgX, -ZnX).

A myriad of methods have been developed, each differing in the required monomer types and catalyst precursors, and to describe the intricacies of these pathways would require a more complete treatise on transition metal-catalyzed cross-coupling reactions than space allows. Instead, the essence of many of these polymerizations can be conveyed using generic terminology. Here, terms such as "catalyst activation", "oxidative addition", "transmetallation", and "reductive elimination" will be used to describe the important steps common among the various catalytic cycles. Accordingly, Scheme 9, which is the accepted mechanism for the Stille coupling reaction [12], will be used as an illustrative diagram highlighting the important components of a given polymerization.

Many of these coupling reactions do not begin with the active catalyst, and thus require some "activation" step as shown in segment A. In a Stille coupling reaction, it is possible to use a palladium(II) catalyst (typically a dihalo compound) which is attacked by two trialkylstannyl-funtionalized monomers, resulting in a diarylpalladium(II) species and halotrialkyltin by-product. Reductive elimination affords the active palladium(0) catalyst and homocoupled biaryl. It is wise to account for catalyst activation when calculating monomer ratios, as the stoichiometric imbalance caused by homocoupling often leads to materials with decreased molecular weights.

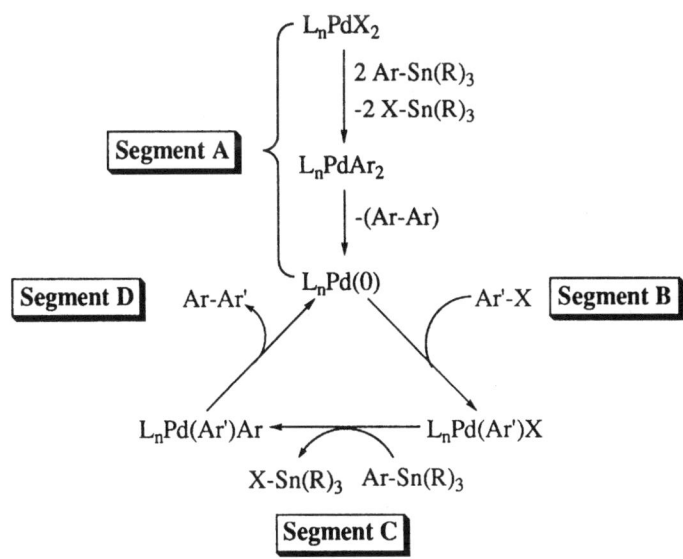

Scheme 9

The actual catalytic cycle is now entered at this point, and oxidative addition of the halo-aryl monomer can occur, as shown in segment B. This step highlights the ability of the active catalyst to insert into aryl-halide bonds, and is so named because the palladium center is now in a 2+ higher oxidation state. The electropositive nature of the aryl-stannyl allows transmetallation to occur (segment C), and as in segment A, a diaryl-metallo species results with the concomitant formation of a small molecule ($XSnR_3$). Finally, the diaryl complex spontaneously undergoes reductive elimination in segment D resulting in C-C bond formation and regeneration of the active catalyst. This cycle is continually repeated, utilizing monomer and oligomer units, ultimately affording the desired conjugated polymer.

As mentioned above, there are a variety of transition metal-catalyzed polymerizations that have been developed. Which coupling method will yield the best results is a function of many factors. For instance, Yamamoto methodology is not feasible with certain functionalities as the required Grignard reagent is extremely reactive [13]. When working with aqueous systems, the Suzuki polymerization is extremely well-suited while most others are not [14]. Moreover, the electronic nature of the monomers can render one method relatively ineffective but increase the effeciency of another. Given these subtle differences, a broad overview of each polymerization route might prove misleading, and hence, we will present examples and generalizations for the different families of CPs in the upcoming sections.

4
Polyacetylene

4.1
Direct Polymerization of Acetylene

Polyacetylene (1) can be viewed as the simplest conjugated polymer and has been studied extensively since the initial reports of high conductivity in doped complexes in 1977 [15, 16]. Polyacetylene serves as a model for developing both the electronic and physical properties of electronically conducting polymers with unique properties in both its reduced semi-conducting and highly doped conducting forms. Polyacetylene is highly crystalline, which has allowed direct structural analysis of both the doped and undoped forms. The chemistry structure and electrical properties of polyacetylene have been reviewed extensively, [17–21] and have also been the subject of several books [22, 23].

1

Structure 1

4.2
Substituted Polyacetylenes

The direct polymerization of substituted acetylenes has been employed extensively in the search for stable and processable PA-based materials. A variety of new monomers and new synthetic chemistry have been developed over the years for the synthesis of processable substituted polyacetylenes [24]. A wide range of synthetic techniques, including thermal, radical, anionic, electrochemical and metathesis polymerizations, have been utilized. The substituents incorporated onto the polyacetylene backbone include phenyl, alkyl, trifluoromethyl, trimethylsilyl, and fused rings, as well as many other functionalities [25, 26]. Acetylene monomers containing mesogens have been synthesized and polymerized in search of highly ordered and oriented polymers [27, 28]. In general, the electrical conductivity of substituted polyacetylenes in their fully doped states is several orders of magnitude lower than that of the unsubstituted parent polymer.

4.3
Cyclic Polyacetylene Derivatives

One of the first methods employed for the preparation of a derivatized polymer having a fully conjugated PA backbone involved the cyclopolymerization of diyne monomers to yield a chain containing fused rings, as shown in Scheme 10 for the hexadiyne example [29, 30].

In this instance, Ziegler-Natta polymerization yields a soluble, linear polymer 2, containing a six-membered cyclic ring fused at each repeat unit. Unfortunately, this polymer undergoes isomerization to form a non-conjugated polymer, disrupting the electronic properties of the backbone [31]. It was found that this isomerization could be prevented by the introduction of heteroatom functionality into the diyne architecture, as exemplified by the polymerization of propiolic anhydride 3, which yielded a stable polymer 4 as shown in Scheme 11 [32].

Scheme 10

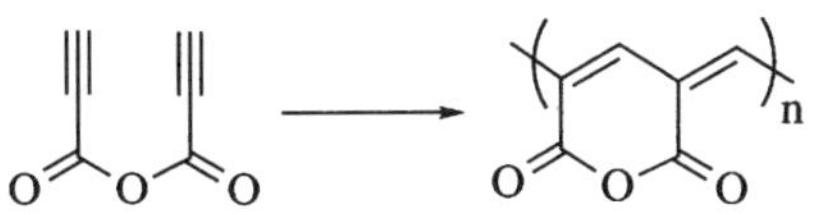

Scheme 11

Further functionalization has been possible using metathetic polymerization routes with the added benefit that high molecular weight and soluble materials can be obtained directly [33, 34]. In these instances, it was found that numerous heteroatoms could replace the anhydride functionality in structure 3. For example, silyl derivatives containing alkyl groups affixed to the silicon atom within the ring afforded a system with appreciable solubility in common organic solvents. One unique copolymer 5 was synthesized having the chemical structure shown below in Scheme 12. The cyano biphenyl-containing polymer exhibited a smectic phase while the methoxy biphenyl-containing polymer exhibited a nematic ordered structure upon heating. Furthermore, I_2 doping of these polymers yielded materials with conductivities in the range of 10^{-2} to 10^{-3} S cm^{-1}.

4.4
Water-Soluble Polyacetylenes

One of the more useful synthetic tricks developed for the preparation of processible conjugated polymers has been the incorporation of ionic pendant groups that can induce water solubility. As polymer chemists look to prepare materials that are more environmentally benign, these water-soluble polymers may ultimately prove to be of commercial importance.

Accordingly, a series of water-soluble PAs were synthesized by the quaternization and subsequent polymerization (via activation by a pyridine nucleus) of 4-ethynylpyridine to yield the substituted PAs 6 shown in Scheme 13 [35–39]. In this thermal polymerization, it was found that alkylation of the pyridine ring was required to ultimately induce polymerization. These quarternized PAs were found to be amorphous solids, and were also reported to be highly soluble in both polar organic solvents and water. Further tailoring of their solubility could be effected through the use of various gegenions.

Spectroscopic and viscosity results for these quarternized PAs indicated that these polymers have effective conjugation lengths ranging from 10–16 double

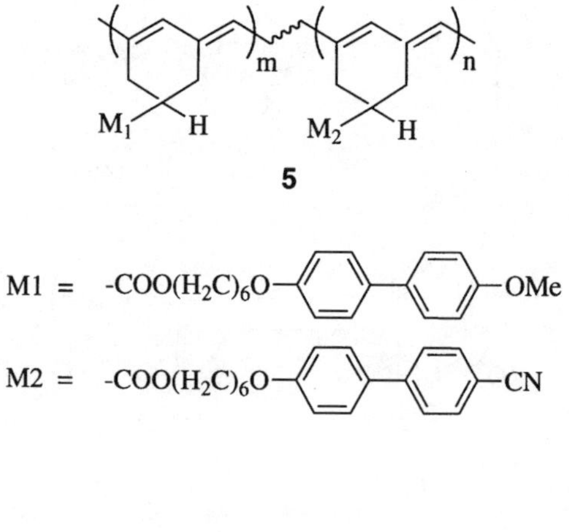

M1 = -COO(H$_2$C)$_6$O-⟨⟩-⟨⟩-OMe

M2 = -COO(H$_2$C)$_6$O-⟨⟩-⟨⟩-CN

Scheme 12

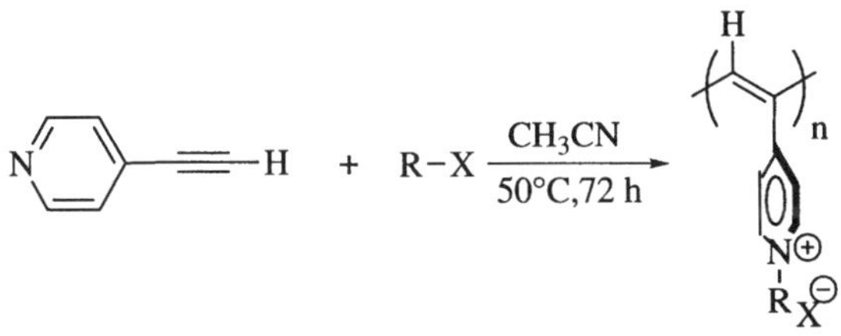

$$\text{N}\underset{}{\overset{}{\diagdown}}\text{—≡—H} + \text{R–X} \xrightarrow[50°C, 72\ h]{CH_3CN} \quad \mathbf{6}$$

Scheme 13

bonds. Here, the effective conjugation length is the average number of double bonds along the chain with effective overlap. This can be limited by either the molecular weight of the polymer (for short chains), or the presence of defects and/or steric distortions that break (or reduce) π-overlap along the backbone. The quarternized polyacetylenes are strongly absorbing in the UV-visible range with absorbances extending out to 700–800 nm.

These quarternized polymers can be viewed as "self-doped" but exhibit relatively low intrinsic conductivities. The polymers can be oxidatively doped with iodine, or reductively doped with TTF, to give highly conducting polyacetylenes with conductivities of 10^{-4} and 10^{-1} S/cm, respectively. One additional attractive feature of this system is that, unlike PA, these quarternized PAs are very stable in air.

4.5
Precursor Polyacetylene Via Ring Opening Metathesis Polymerization (ROMP)

The precursor polymer route has been developed as one of the most important methods for the preparation of conjugated polymers [40, 41]. As the fully conjugated PA systems tend to be completely intractable, the synthesis of a soluble precursor allows processing prior to conversion to the fully conjugated form. In addition, as the polymer retains solubility during synthesis, high molecular weights are easily obtained. Numerous precursor PAs have been prepared via the metathesis polymerization of cyclobutene derivatives, as illustrated in Schemes 14 and 15. Thermolysis of these materials leads to an evolution of a small organic byproduct, driving the formation of the conjugated backbone. A nice feature of this method is that the prepolymer can be oriented by stretching during thermal elimination. Furthermore, the gaseous by-product serves as a plasticizer, which facilitates ordering of the chains during orientation, ultimately leading to high conductivities in the doped state. A drawback of this precursor method is the relatively high reactivity of the material (explosive) when converting large amounts of material to polyacetylene [40, 41]. Such behavior was especially seen for polymer 7 and motivated the synthesis of the less reactive derivatives **8–11**.

The ROMP methodology has been used extensively for the further synthesis of other precursor polymers and substituted PAs. Cyclic rings containing one or

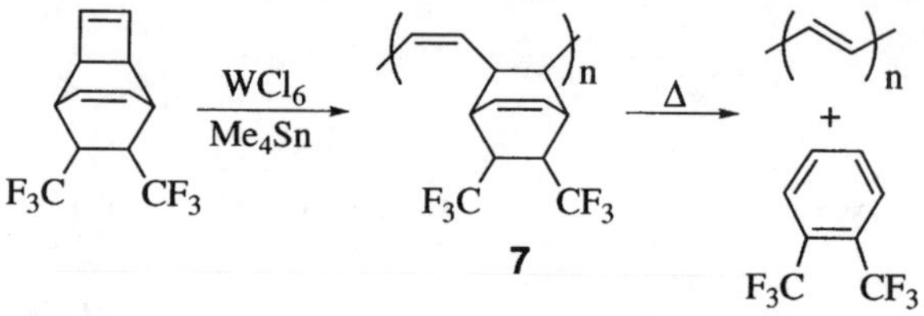

Scheme 14

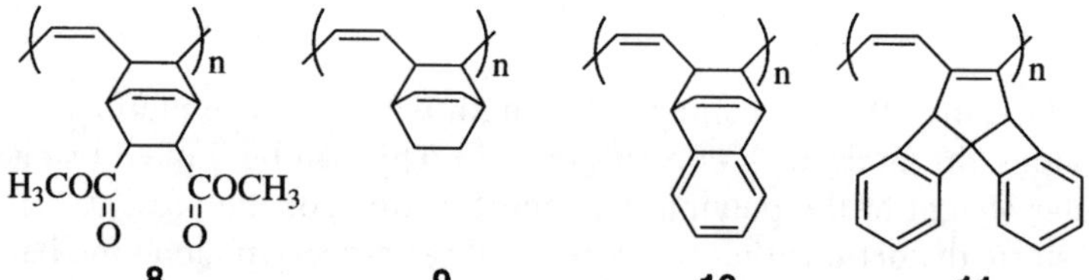

Scheme 15

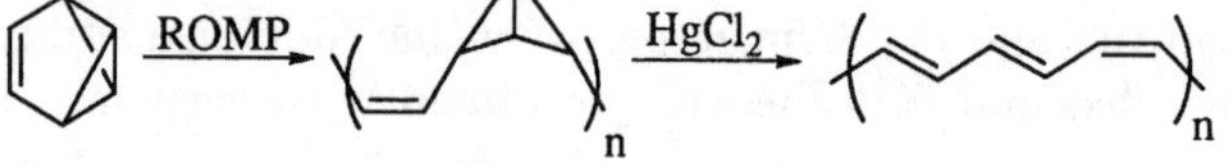

Scheme 16

more double bonds can be polymerized by the many known metathesis catalysts. Details of the different types of ROMP catalysts available, their synthesis and mechanisms of polymerization, have been reviewed [10].

The flexibility of the ROMP methodology has allowed numerous systems to be investigated, as both polymer precursors and conjugated PAs can be synthesized directly. The precursor polybenzvalene **12**, shown in Scheme 16, was the first of the PA-precursors to be synthesized by ROMP that did not require elimination of a molecule during conversion to the conjugated form [42]. The precursor polymer was then converted to *trans*-PA by treatment with zinc iodide or mercuric chloride.

Sequential addition methods were utilized to synthesize block copolymers of polybenzvalene with polynorbornene to yield block copolymers of PA and polynorbornene after isomerization [43]. (It should be noted that both benzvalene and polybenzvalenes are sensitive to shock and mechanical stress causing ring strain-promoted explosions in the precursor materials.)

A broad series of substituted-PAs **13** have been synthesized from cyclooctatetraene (COT) derivatives as a result of the tendency of the COT nucleus to un-

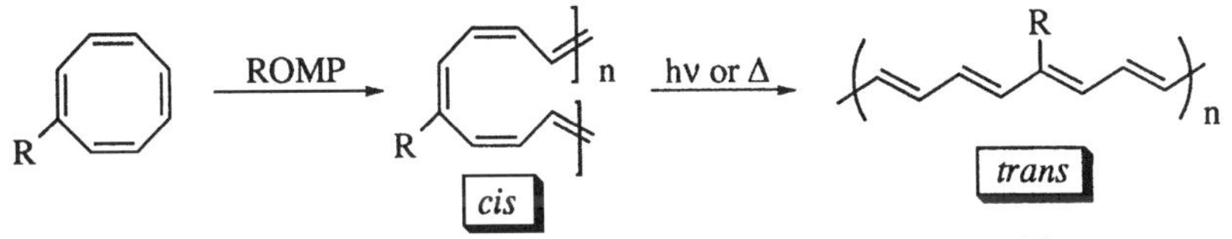

Molecular Engineering of π-Conjugated Polymers

R = n-, s-, and t-butyl, n-octyl, n-octadecyl, neopentyl, 2-ethylhexyl, t-butoxy, isopropyl, cyclopentyl, cyclopropyl, phenyl

Scheme 17

dergo ROMP as shown in Scheme 17 [44–49]. It was determined that monosubstituted monomers polymerized much more effectively than their di- and trisubstituted counterparts due to steric effects of the bulkier rings. In this instance, ROMP polymerizations yielded high molecular weight polymers comprised of predominantly *cis* units and a high level of polydispersity. The polydispersity of these materials has been attributed to a backbiting mechanism, which is not a problem for many of the other ROMP polymerizations. This undesired reaction results in the formation of various substituted-benzene side products. While this side reaction does not terminate the polymerization, it does reduce the overall molecular weight of the polymer and complicates its resulting structure through a decrease in side chain concentration.

As noted above, these polymerized COTs yield conjugated polymers with a high level of *cis*-isomer content, affording a higher level of solubility in the substituted polymers. Electronic spectroscopy suggests that the polyCOTs have effective conjugation lengths of 15 to 20 double bonds. It was found that the electrical conductivity of these materials was dependent upon the *cis*- and *trans*-isomer content in the polymer. Isomerization to the more conducting *trans*-form can be accomplished via either photochemical or thermal means, but results in a concomitant reduction in the polymer's solubility.

5
Poly(*p*-phenylene)

Poly(*p*-phenylene) (PPP) **14** is a highly crystalline, infusible, and insoluble polymer that has an exceptional degree of thermal stability in its neutral form [1a, 50]. While there has been a significant amount of work published throughout the years concerning the synthesis of PPPs, a review of the literature indicates that, in many early cases, the polymers' structures could not be well-characterized due to their insolubility [51–54]. In more recent times, synthetic methodologies have been developed that now allow PPPs to be prepared that are structurally regular and processible, and these more amenable materials are the main focus of this section.

5.1
Direct Synthesis of Poly(p-phenylenes)

The direct synthesis of PPP by condensation polymerization represents one of the oldest preparations of a conjugated polymer. This is exemplified by the Ullmann coupling of dihalobenzenes using copper powder [56] and by Wurtz [57, 58] chemistry where dihalobenzenes are reacted with sodium metal to give oligomeric PPPs [59].

The synthesis of PPP via step polycondensation Friedel-Crafts polymerization of benzene was developed by Kovacic, as outlined in Scheme 18 [1a, 60–62]. Lewis acids, including $AlCl_3$, $AlBr_3$, $FeCl_3$, and $SbCl_5$, (along with an oxidizing agent) are used as the initiator system for this style of polymerization. A review of this polymerization shows that both cross-linked and branched polymers can be obtained depending on reaction conditions and the nature of the polymerization agent.

An important advancement for the synthesis of well-defined PPPs was the site-controlled nickel-catalyzed coupling of Grignard reagents, derived from dihalobenzenes, to yield linear PPP as shown in Scheme 19 [63, 64]. The PPPs obtained by Yamamoto chemistry range from yellow to brown infusible solids of low molecular weights; typically with degrees of polymerization ranging from 10–12.

5.2
Soluble Substituted PPPs

The incorporation of pendant alkyl substituents onto the PPP backbone allows the material to be solubilized in common solvents and greatly enhances processibility. However, a resulting trade-off in electrical properties can often arise due to the steric bulk of the solubilizing groups. In many cases the twist angle between consecutive rings (23° for unsubstituted PPP) [65] significantly increases, reducing π-electronic overlap and broadening the band gap of the material. Soluble PPPs 15 were first prepared by the incorporation of long alkyl side chains at

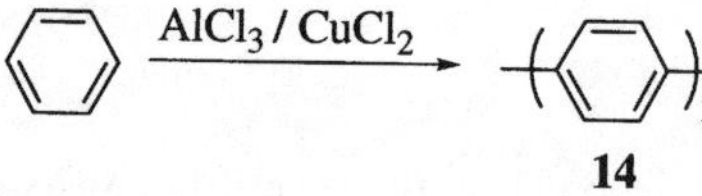

Scheme 18

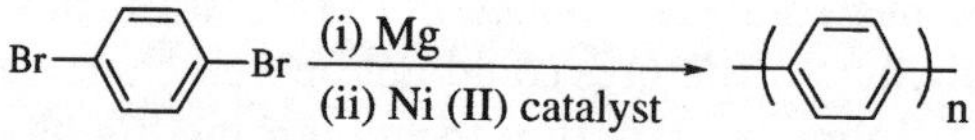

Scheme 19

every repeat unit as shown in Scheme 20 [66]. The solubility of the polymers was found to be a function of the length of the side chain, with alkyl groups longer than 6 carbons, yet shorter than 12, giving an optimum balance for physical and electronic properties. An average degree of polymerization of 8–20 units was found for these polymers.

Suzuki coupling chemistry of benzene boronic acid derivatives and halobenzenes using a Pd(0) catalyst has also been employed for the synthesis of substituted PPPs as illustrated by the A-B type monomer **16** [67–73]. These initial syntheses were carried out under heterogeneous conditions at a basic pH as illustrated by Scheme 21. Such Suzuki coupling polymerizations are rather attractive alternatives as a wide variety of functional groups can be tolerated with minimal interference in the coupling scheme.

Recently, a number of studies have been aimed at expanding the scope of palladium- and nickel-catalyzed polycondensation reactions through new twists on established routes or the advent of entirely new reaction pathways. The Ni(0)-catalyzed homo-coupling polymerization of triflate-substituted benzenes **17**, as shown in Scheme 22, was utilized to synthesize a number of PPP derivatives [74–76]. Structural characterization indicated that these polymers are *para*-linked,

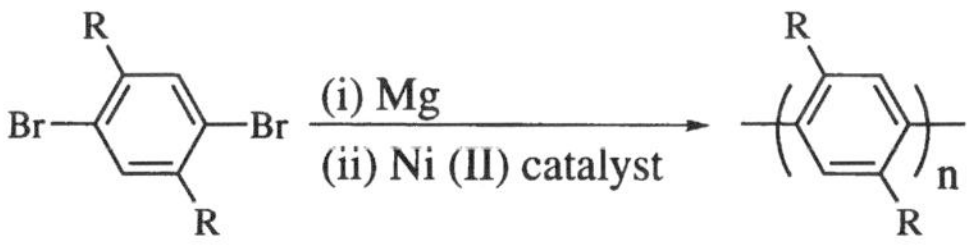

Scheme 20

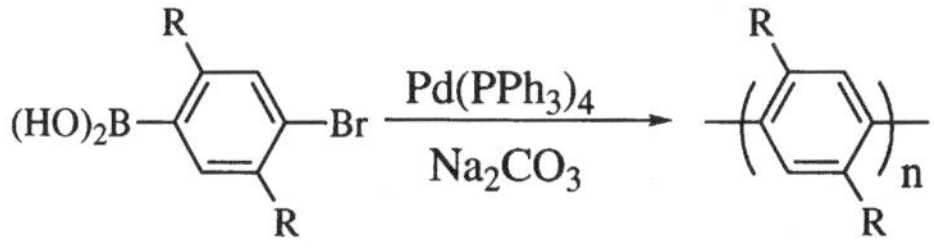

Scheme 21

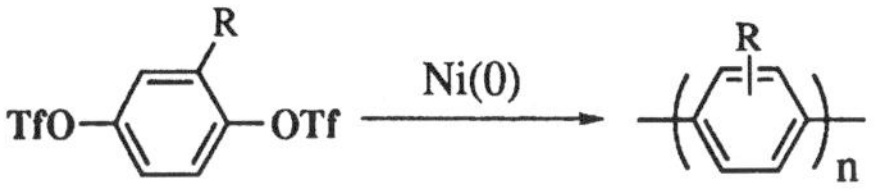

Scheme 22

R = *t*-Butyl, Ph, CO_2CH_3

but that the pendant groups at the ortho-positions can undergo isomerization in the presence of the catalyst to yield polymers with randomly distributed substituents along the backbone. Phenyl- and *t*-butyl-substituted monomers gave poor yields of low molecular weight polymers as steric hindrance reduced reactivity towards polymerization (relative to the use of non-sterically interacting substituents such as methoxy groups). Accordingly, the number average molecular weights of these polymers ranged from 1000–6000 g/mol depending on the steric nature of the substituent and polymerization conditions used.

A Pd(0)-catalyzed cross-coupling polymerization of aryl triflates and aryl stannanes has also been reported allowing a variety of phenylene copolymers to be produced [77]. In an extension of the triflate-coupling work, analogous methodology was developed to synthesize PPPs from aryl mesylates [78–81]. A major benfit of this coupling chemistry is that the use of mesylates circumvents the costs associated with preparing expensive triflate monomers. Furthermore, this polymerization methodology also exhibits the ability (employing certain conditions and monomer architectures) to produce PPPs with much higher molecular weights than materials produced by the corresponding triflate pathway.

A number of polyphenylenes have been synthesized by routes employing a zero-valent Ni-catalyzed coupling previously utilized for biaryl synthesis [82–87]. For instance, relatively high molecular weight poly(benzoyl-1,4-phenylene) **18**, shown below in Scheme 23, was synthesized via the homopolymerization of 2,5-dichlorobenzophenone. It was shown that polymer properties were largely dependent upon the polymerization conditions (coligand, temperature, and reaction time) with molecular weights ranging from 4400–8900 g/mol being obtained for these polymers.

5.3
Soluble, Fully Aromatic Polyphenylenes

The synthesis of soluble, fully aromatic-substituted polyphenylenes **19** has been pursued, as outlined in Scheme 24, where polymerization of various dihalobenzenes was accomplished by initial activation with *t*-butyl lithium [88, 89]. In this example, the use of 1.5 equivalents of *t*-butyl lithium in conjunction with dibromobenzene in THF at –78 °C, followed by warming to room temperature, yielded **19**.

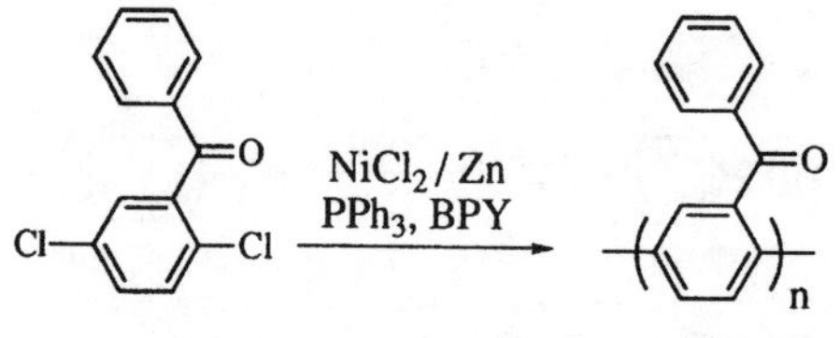

Scheme 23

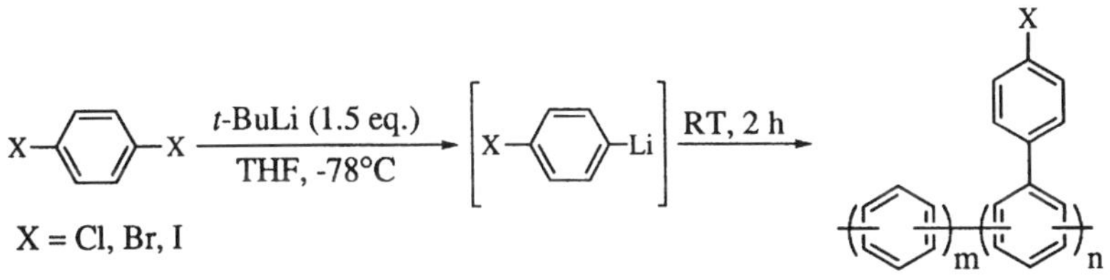

19

Scheme 24

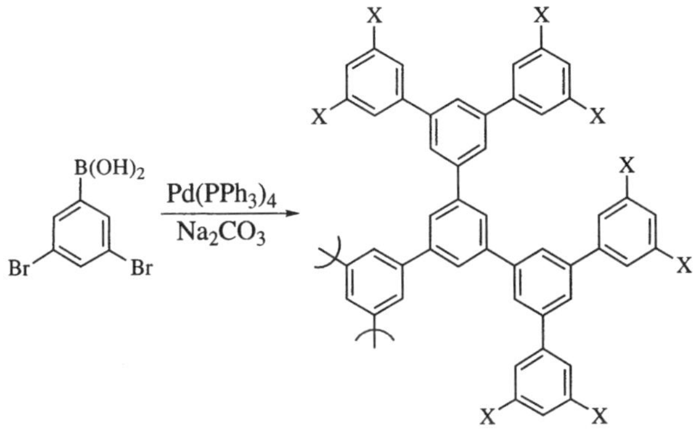

20a X = Br (1) BuLi (2) CO_2

20b X = -COOLi

Scheme 25

The mechanism for this polymerization is believed to involve the formation of an *o*-benzyne intermediate that gives rise to either *meta*- or *para*-linkages along the polymer backbone. Structural investigations have indicated the resulting organic-soluble polymer does indeed contain both *meta*- and *para*-linkages with X-ray diffraction studies showing the polymer to be rather amorphous in nature.

Novel, hyper-branched polyphenylenes **20** have been prepared by the self-coupling of 3,5-dibromophenylboronic acid in the presence of $Pd(PPh_3)_4$ as shown in Scheme 25 [90, 91]. (The monomer was prepared by treatment of the monolithiate of 1,3,5-triboromobenzene with trimethyl borate.) The hyper-branched polymer was found to be organic soluble, and could be converted to a water-soluble, polyphenylene derivative by treatment with butyl lithium followed by quenching with CO_2. The water-soluble nature of such hyper-branched polyphenylenes has made them suitable candidates for use in various applications such as unimolecular micelles.

5.4
Polyphenylenes Via Soluble Precursor Polymers

As with polyacetylenes, high molecular weight PPPs can be prepared using the soluble precursor method. Early work demonstrated that 1,3-cyclohexadiene could be polymerized by anionic [92], cationic,[93, 94], and Ziegler-Natta [93] polymerization routes. The polycyclohexadienes obtained were converted to a π-conjugated polymer by dehydrogenation under a variety of conditions. One significant drawback to this approach is the fact that these polymerization routes afford materials with a significant defect content due to 1,2-linkages contained in the prepolymer and incomplete dehydrogenation/aromatization.

The successful synthesis of a high molecular weight precursor to polyphenylene that could be more readily converted to its corresponding conjugated polymer was reported where the prepolymer utilized ester substituents [95, 96]. In a novel bacterial oxidation of benzene, a *cis*-cyclohexadiene diol **21** was initially prepared that was later acetylated and polymerized as shown in Scheme 26. This polymer was determined to contain approximately 90% 1,4-linkages and 10% 1,2-linkages.

Subsequently, a non-enzymatic synthesis of the *cis*-diesters of cyclohexadiene was developed. Radical polymerization of these esters again yielded polymers with mixed 1,4- and 1,2-linkages [97]. Despite the significant defect level of these polymers (1,2-linkages), the PPPs synthesized by these routes exhibited electrical conductivities between 1 and 100 S cm^{-1} upon oxidative doping.

A modified synthesis of the above precursor polymer has been reported using a TMS derivative of the cyclohexadienediol [98]. This bisTMS derivative was then polymerized using bis[(allyl)trifluoroacetatonickel(II)] (ANiTFA)$_2$, as shown in Scheme 27, to give exclusively 1,4-linked polycyclohexadiene **22** boasting a rather high molecular weight. While this precursor polymer did not undergo direct elimination due to the poor nature of the TMS leaving group, it was converted to the diol-containing polymer by treatment with fluoride ion. Esterification of the resulting polydiol yielded an ester-functionalized polymer that eliminated cleanly at 320–340 °C to yield PPP.

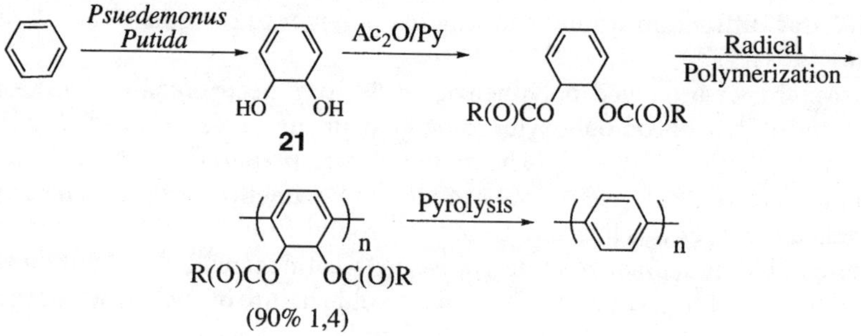

Scheme 26

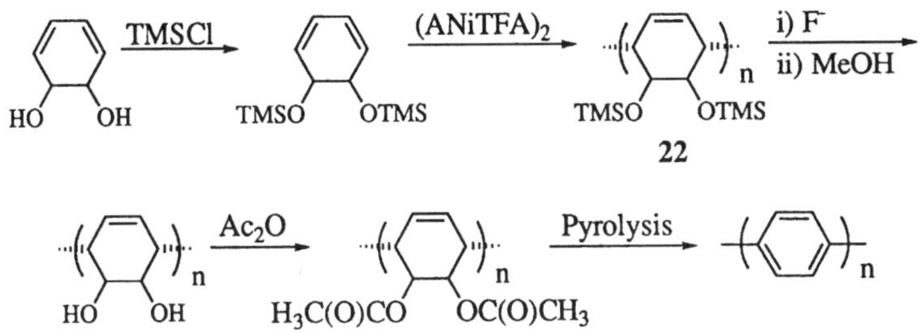

Scheme 27

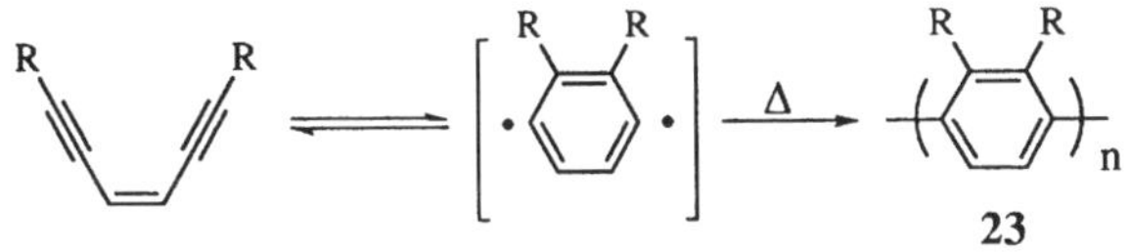

Scheme 28

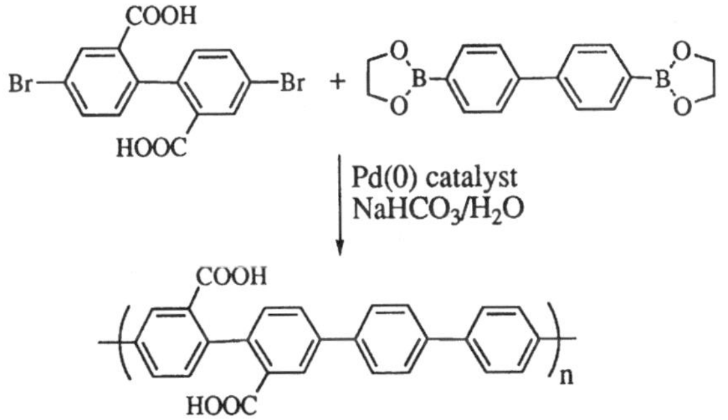

Scheme 29

A Bergman-like [99, 100] cyclization has been employed to synthesize substituted-PPPs **23** [101]. Using this strategy enediynes are cyclized and subsequently coupled upon thermal treatment, as shown in Scheme 28. The polymers obtained in this manner are 2,3-disubstituted and display number average molecular weights on the order of 1500–2500 g/mol.

5.5
Water-Soluble Poly(p-phenylenes)

Water-soluble poly(p-phenylene) **24**, shown in Scheme 29, was prepared by the introduction of carboxylic acid pendant substituents along the p-phenylene chains [102]. In initial work in this area, a dicarboxy-substituted dibromobiphenyl was polymerized with 4,4'-biphenyl bis-boronic acid via Suzuki coupling

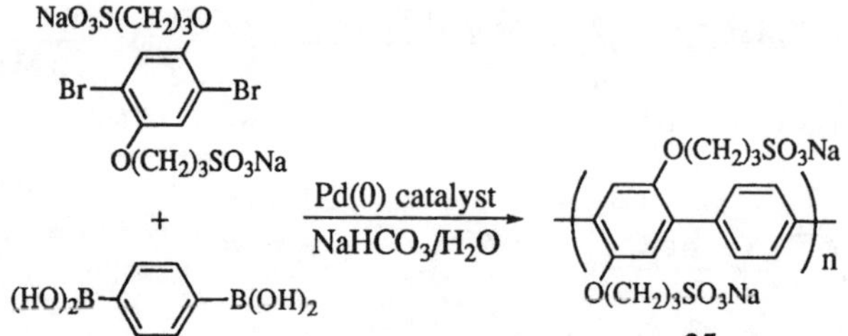

Scheme 30

chemistry in the presence of a water-soluble Pd(0) catalyst. A 0.1% solution of the polymer forms highly viscous solutions in 0.1 M aqueous Na_2CO_3, and thin transparent films obtained from solution casting were discovered to be birefringent under polarized light.

The synthesis of water-soluble PPPs **25** containing propoxysulfonate groups, shown in Scheme 30, has been accomplished by the polymerization of 1,4-dibromo-2,5-bis(3-sulfonatopropoxy)benzene with benzene bisboronic acid as outlined in reaction 24 [103]. Using such methodology, polymers could be obtained displaying a wide range of PPP chain lengths (from low molecular weight oligomers to chains greater than approximately 50 rings) by utilizing a monofunctional comonomer. This polymer was highly soluble in water and could be solution cast to give both p- and n-dopable films exhibiting optical band gaps of 3.0 eV (Scheme 30).

5.6
"Planarized" Polyphenylenes

As has been outlined in this section, unsubstituted-PPP is an insoluble, infusible material. Substitution of the polymer backbone with appropriate alkyl- or alkoxy-groups can, in turn, afford soluble materials. However, in many instances these solubilizing groups prove detrimental to electronic properties as a result of steric interactions with the phenylene units. Accordingly, a recent thrust in PPP synthesis has been aimed at obtaining electronically optimized PPPs through the use of "planarized" structures. This strategy involves placing bridging units that will hold consecutive aryl rings in a near-planar geometry. In addition, the use of conjugated connectors can help to provide even more π-conjugation than possible in the parent PPP.

Utilizing Suzuki polymerization methodology, ladder-type PPPs **26** that were both soluble and structurally well-defined have been prepared [104–109]. The synthesis of this elegant polymer is illustrated in Scheme 31 with the uncyclized-PPP being obtained via a Suzuki coupling polymerization. The key step in this scheme is the final bridging of the phenylene rings, which was accomplished in this example by a Friedel-Crafts alkylation/cyclization. Such methodology must

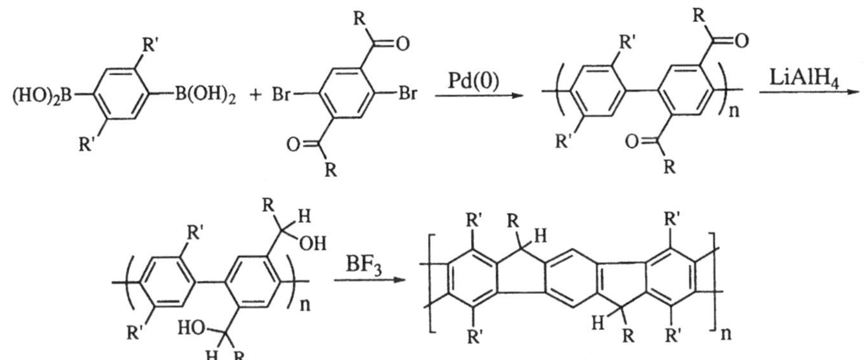

Molecular Engineering of π-Conjugated Polymers

Scheme 31

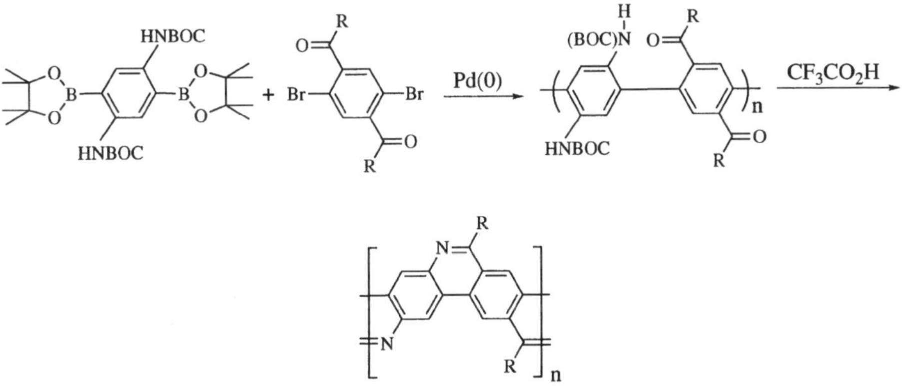

Scheme 32

be essentially quantitative in conversion to afford a highly regular structure in the ultimate ladder polymer. The electronic benefits of the planarized structure can be seen in the UV-Vis absorption spectrum for the polymer. The solution l_{max} for the polymer occurs at 438 nm, quite close to the theoretically predicted value of 442 nm for regular, fully conjugated unsubstituted-PPP [104].

Further use of the Suzuki polymerization has yielded polymeric precursors to planar PPP derivatives **27** [105, 106] (Tour '93 and Lamba). The strategy, shown above in Scheme 32, utilized a postpolymerization cyclocondensation reaction to afford imine-bridged aryl units. Optical absorption data showed that the planarized polymers exhibited large bathochromic shifts relative to the uncyclized parent polymers.

6
Poly(phenylene vinylene)

6.1
Introduction

Poly(phenylene vinylene) (PPV) can be viewed as an alternating copolymer composed of *p*-phenylene and *trans*-vinylene units [112]. The optical band gap of unsubstituted-PPV is reported to be ~2.6 eV, which is intermediate between those of PPP and polyacetylene [113–115]. This material has consistently been one of the most-studied conjugated polymers as doped PPV films have electrical conductivities in the range of 10–500 S cm^{-1} [116, 117], and exhibit large third-order nonlinear optical responses [118-121]. With their typically bright fluorescence, PPV and its derivatives have become one of the most widely employed family of materials for polymer-based LEDs [122-130]. For example, dialkoxy-substitution of the backbone's phenylene rings results in a strong increase in the corresponding polymer's photo current, important for solar energy conversion applications [131]. Furthermore, the *trans* form of PPV is an extremely rigid molecule capable of dense packing and many of its copolymers have exhibited liquid crystallinity. Accordingly, thermotropic liquid crystalline behavior in a number of PPV derivatives has been reported, and several types of structural modifications have been made to lower the softening or melting point of the material [132, 133].

The synthetic chemistry of PPV has consistently improved over the years. Original polymerizations used common olefin-forming reactions that tended to produce low molecular weight, oligomeric materials. Significant breakthroughs have surfaced, and it is now possible to synthesize high molecular weight, processable polymers through improved chemical synthesis. Of these new synthetic pathways, metathesis polymerization, Heck's reaction and bissulfonium salts are the most extensively studied reactions for the preparation of substituted- and unsubstituted-PPVs.

6.2
PPV Via Classical Organic Means

Initially, PPV was synthesized by condensation reactions that can be found in most introductory organic textbooks. Most commonly, reactions such as the Wittig, Knoevenagel, Wurtz-Wittig, or McMurray, as well various dehydrohalogenation reactions, were employed that yielded oligomers or low molecular weight PPV.

A "Wittig" style polymerization, shown in Scheme 33, is the result of condensation of dialdehyde monomers with bis(phosphonium) salts containing aromatic cores, and was reported for the first time in 1960 [134]. Unfortunately, due to low reactivity and conversion, the Wittig polymerization typically only affords materials with a DP of ~10. Despite its limitation to forming low molecular weight polymers, this chemistry has been used to make a variety of soluble and insoluble PPVs **28** [134–139].

The Knoevenagel reaction is a base-catalyzed condensation between a dialdehyde and an arene possessing two sites with relatively acidic protons. In this polymerization, shown schematically in Scheme 34, deprotonation affords a difunctional nucleophile that subsequently attacks the carbonyl functionalities present in the other monomers. Elimination is the final step in the Knoevenagel sequence, and the use of monomers with highly acidic protons drives the reaction to completion. A number of research groups have employed this method to obtain PPV and its substituted analogs **29** [126, 140–146].

Cyano-substituted PPVs have low electrical conductivity due to the electron withdrawing ability of cyano group, however, they have become important materials for the fabrication of LEDs. While LED devices constructed from unsubstituted-PPV exhibit efficiencies of less than 1%, similar LED devices fabricated with cyano-substituted PPVs have displayed efficiencies of over 4%. To further enhance device performance, proccessible polymers containing cyano functionalities were synthesized utilizing monomers possessing solubilizing alkoxy chains for ease of device construction.

Dehydrohalogenation reactions have also been employed for the synthesis of a variety of PPVs. Polymerization in this manner is generally carried out using a variety of strong bases in conjunction with a,a'-dihaloxylenes [147–149]. In Scheme 35, a,a'-dichloroxylene is subjected to dehydrohalogenation conditions using NaH with a DMF solvent system to give unsubstituted-PPV **30**. Polymers prepared by dehydrohalogenation have been known to contain significant amounts of residual chlorine, and heating of the polymer to 300 °C helps to complete elimination [138].

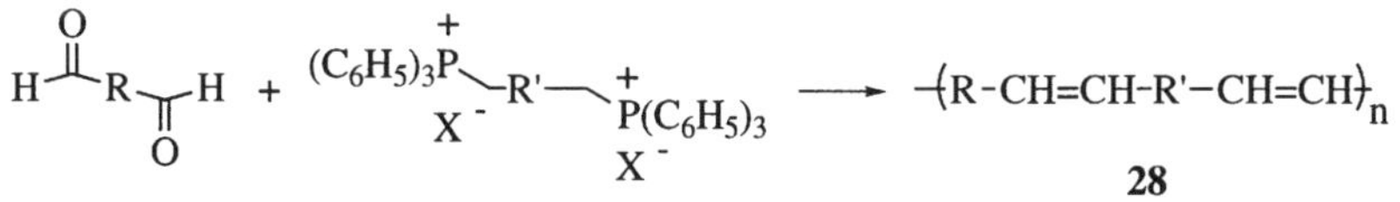

Scheme 33

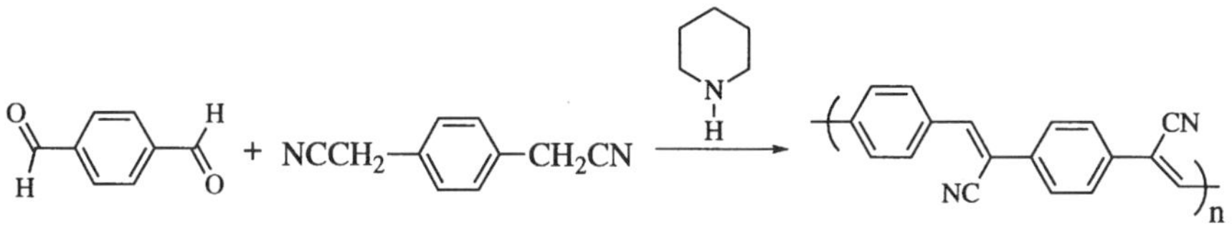

Scheme 34

Scheme 35

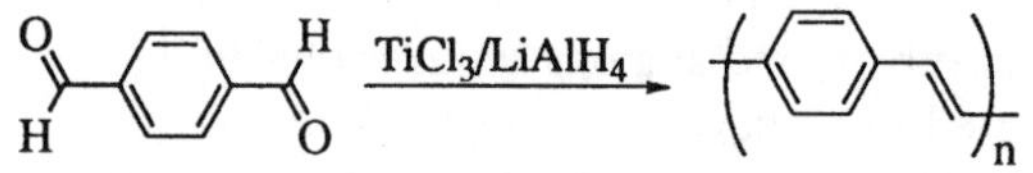

Scheme 36

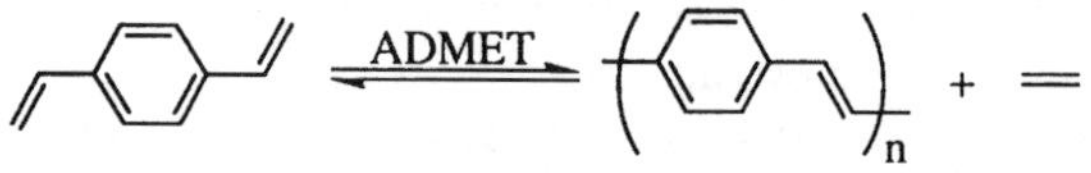

Scheme 37

The McMurray reaction, shown above in Scheme 36, has also been called upon for the synthesis of PPV [150–152]. Here, deoxygenative coupling of a dialdehyde is accomplished by reduction of the carbonyl groups with Ti(0) (generated by the reduction of $TiCl_3$ with $LiAlH_4$). The formation of TiO_2 drives the polymerization and ensures the removal of both carbonyl oxygens.

6.3
Direct Synthesis of PPV via ADMET

Acyclic diene metathesis (ADMET) polymerization of divinyl benzene, shown in Scheme 37, using an extremely reactive tungsten alkylidene catalyst (Schrock's catalyst) yielded PPV oligomers with DP of ~8 [153]. In this example ethylene is formed as a side product of metathesis, and its removal by the use of high vacuum helps drive the polymerization in the forward direction.

Soluble PPV derivatives can be prepared by ADMET polymerization through the use of dialkyl divinyl benzene monomers [154, 155]. Accordingly, 2,5-diheptyl-1,4-divinylbenzene was polymerized using Schrock's catalyst yielding monodisperse polymer that was readily soluble in THF and $CHCl_3$. ^{1}H NMR and GPC results for this sample suggested a DP of ~10. Unfortunately, polar groups have tended to inhibit metathesis and to date soluble PPV produced via ADMET is limited to monomers possessing alkyl substituents.

6.4
Precursor Polymer Route

This polymerization strategy is perhaps the most widely employed for the synthesis of PPV and its substituted derivatives, as well as various other heteroaromatic vinylenes. While there are a number of approaches that fit into this category, they all proceed through a soluble "pre-polymer". This pre-polymer is treated to effect elimination ultimately yielding the desired PPV derivative. The materials produced by this method can be of very high molecular weight, and their films highly oriented by stretching during conversion of the precursor polymer to its conjugated form.

The Wessling and Zimmerman aqueous precursor route is illustrated in Scheme 38 [156]. Here, a bis(halomethyl)monomer is reacted with dimethylsulfide and subsequent treatment with base affords the high molecular weight precursor polyelectrolyte 31. Due to the instability of 31, polymerization must be carried out at low temperatures (<4 °C) to avoid thermal elimination of the polyelectrolyte. Precursor polymer 31 can be stored in solution with refrigeration, and its shelf life can be increased by the addition of a small amount of pyridine. Precursor polymer 31 can be processed into highly oriented, free-standing films or fibers that can subsequently be converted to PPV with the elimination of gaseous dimethylsulfide and HCl at 200 °C.

The effect of the structure of the sulfonium salt has been investigated through the use of various sulfides. As shown in Scheme 39, monomer salts 32 and 33

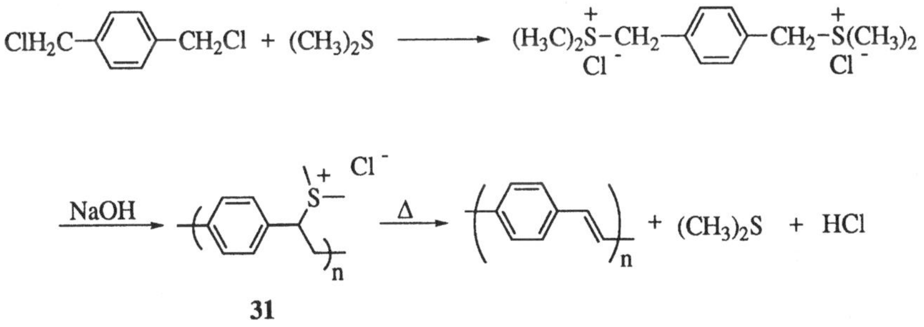

Scheme 38

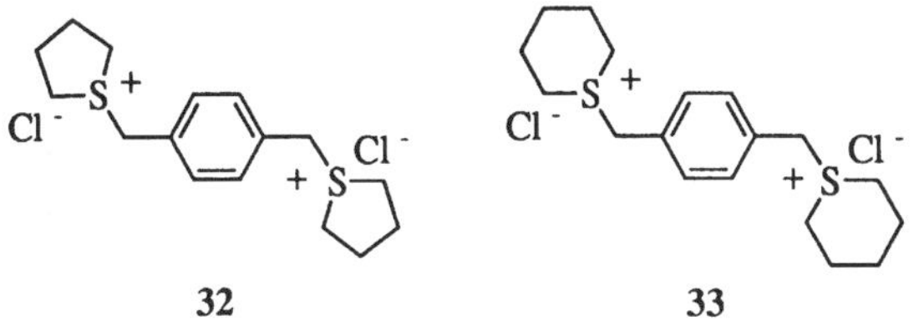

Scheme 39

were synthesized utilizing tetrahydrothiophene and other cyclic sulfides, respectively, in place of dimethyl sulfide [157, 158]. Tetrahydrothiophene salts exhibited the optimum balance between stability of the pre-polymer at low temperatures and ease of conversion to PPV. Most interestingly, it was discovered that PPV obtained from tetrahydrothienyl sulfonium salts displayed better conductivity when compared to polymer obtained using the other sulfonium salts.

The synthesis of PPV was slightly modified using tetrahydrothienyl sulfonium salt-based pre-polymers [159]. The precursor polymer 34 was dissolved in methanol and heated at 55 °C for 18 h to give the methoxy-substituted polymer 35, shown in Scheme 40. The great benefit of this method is that the fully methoxy-substituted polymer 35 is soluble in organic solvents allowing easier processing throught the use of more volatile solvents. Elimination to PPV is accomplished by heating at 220 °C in the presence of HCl gas under an argon atmosphere for 22 h.

The effect of different gegenions on the elimination temperature of the prepolymer and its subsequent manifestation in PPV properties was also investigated [160–162]. As shown in Scheme 41, the Cl⁻ ions of precursor polymer 34 were replaced by F⁻, Br⁻, I⁻, or acetate ions through ion-exchange via dialysis to give the corresponding precursor polymers 36.

Precursor polymer 36d, with acetate gegenions, required significantly higher elimination temperatures than did the other polymers in the series. Thermal

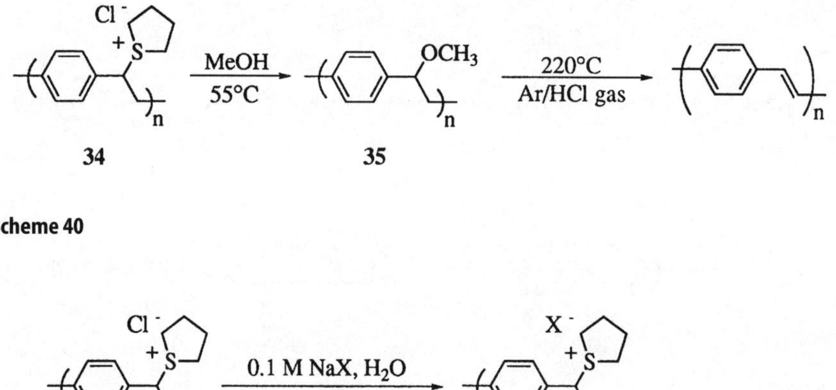

Scheme 40

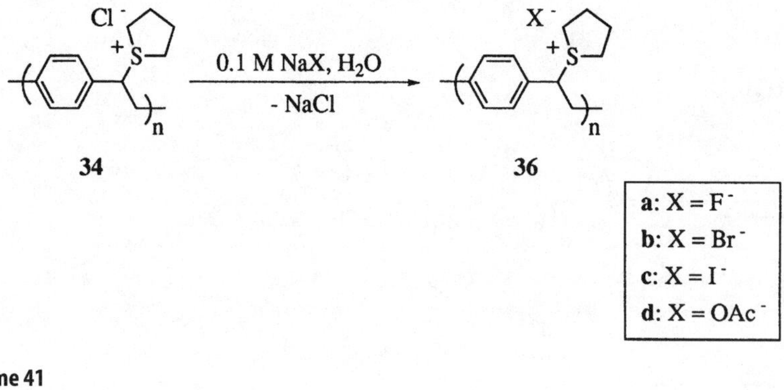

Scheme 41

elimination to PPV is believed to occur through an E2 mechanism. Thermal analysis and IR spectra indicated that water molecules present in the precursor polymer are strongly hydrogen bonded, and the presence of this hydrogen bonded water is essential for lowering the elimination temperature. Accordingly, acetate ions are more hydrophobic than halide anions, reducing the hydrogen bonding of water by the precursor polymer, which is subsequently responsible for the higher elimination temperature.

Various alkoxy-functionalized PPV derivatives have also been synthesized using the bis(sulfonium salt) pre-polymer route. In terms of properties, the presence of the electron donating alkoxy group results in a reduction of the band gap, higher electrical conductivities upon doping, and improved solubility in organic solvents when the side chains are of sufficient length. For instance, dimethoxy-substituted PPV 37 has a band gap more than 0.3 eV less than that of unsubstituted-PPV. While PPV cannot be doped with I_2 due to the polymer's high oxidation potential, alkoxy-substituted PPV derivatives are readily doped with I_2 to give highly conducting polymer films [120, 163–167]. Dihexyloxy-PPV is fully soluble in organic solvents. Conversion of its corresponding precursor polymer to PPV can be performed in basic solution as opposed to thermal treatment in solid state.

Water-soluble, propoxysulfonated-PPV 38 has been prepared as outlined in Scheme 42 [168]. Polymerization, using the bis(sulfonium salt) method, pro-

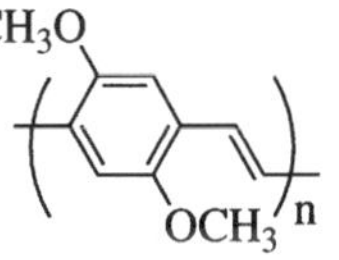

Structure 37

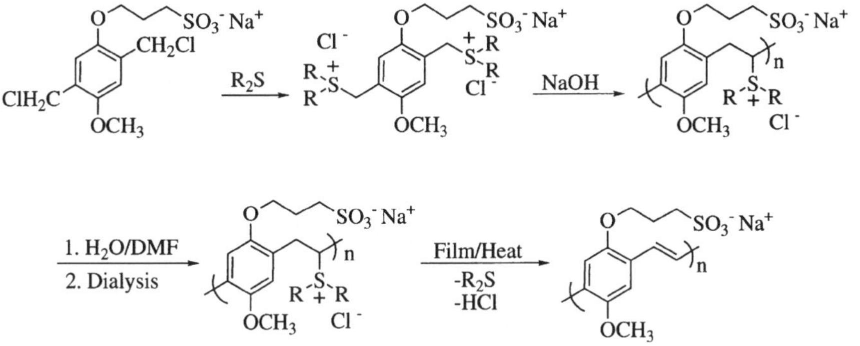

Scheme 42

ceeded smoothly to yield a precursor polymer that could be converted to PPV by a number of methods. Upon elimination, polymer films can be cast from H_2O and doped with HCl. Not only do the sulfonate groups afford water solubility, but they also act as the charge compensating dopant ions during *p*-doping (hence, the polymer can be said to be "self doping"). Conductivity of doped polymer films typically approach 2×10^{-6} S cm^{-1} in air and increases to 10^{-4} to 10^{-2} S cm^{-1} in a humid atmosphere.

Some accompanying disadvantages do exist with synthesizing PPV through the various bis(sulfonium salt) methods. Chiefly speaking, yields of PPV are relatively low (typicaly ranging from 10–20%) due to the dialysis isolation methodology employed. The poor stability of the precursor polymer, coupled with its short shelf life, is also of prime concern. Polymerization of sulfonium salts often generates free alcohol functionalities when performed under basic conditions, and these groups tend to undergo oxidation to carbonyl groups upon thermal elimination, ultimately decreasing the effective conjugation length of the PPV [116]. However, through continual efforts toward enhanced synthetic procedures, many of these difficulties have been overcome.

Precursor routes, other than the bis(sulfonium salt), have also been explored for the synthesis of PPV. For example, PPV was synthesized using sulfoxide and sulfone analogs, and the detailed syntheses of these intermediates are outlined in Scheme 43 [169, 170]. While monomers **39** and **40** are insoluble in water, they

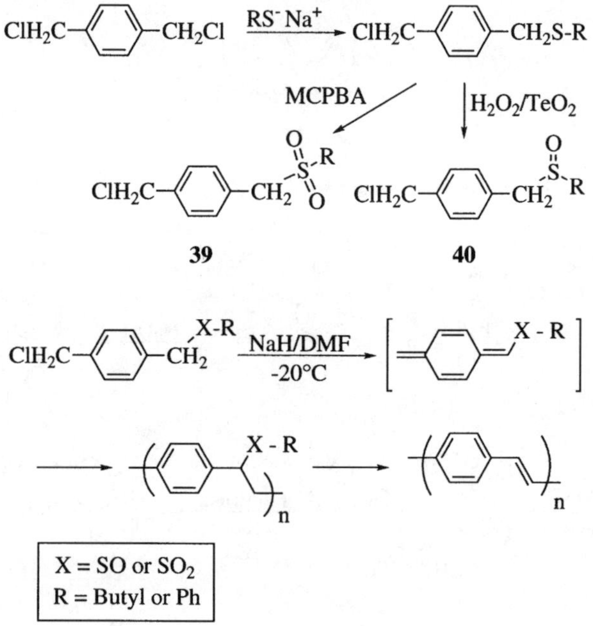

Scheme 43

are soluble in organic solvents, and thus, allow polymerization in organic media (e.g., a NaH/DMF mixture). A very attactive feature of these new strategies is that neither the sulfoxide nor the sulfone precursor polymers are polyelectrolytes in nature, allowing for facile purification via reprecipitation from organic solvents.

The synthesis and characterization of precursor PPVs using ROMP have been reported using substituted bicyclo(2.2.2) octadienes [171–173]. Polymerization of monomer **41**, shown in Scheme 44, can be accomplished in high yield using either a tungsten- or molybdenum-alkylidene catalyst. While the initiation process is rather slow, the reaction proceeds to completion with precursor polymer **42** being obtained with relatively low molecular weight dispersities (PDI~1.2–1.3). Structural analysis of the polymer shows that ROMP, in this case, yields exclusively 1,4-linkages with spectral data suggesting an equal distribution of *cis*- and *trans*-vinylene units.

Thermal elimination at a temperature of 260 °C resulted in a yellow PPV film, and spectral characterization of the polymer indicated the aromatization reaction was both complete and possessed fewer defects than the analogously prepared PPV. Interestingly, this elimination temperature can be lowered by ca. 60 °C through the addition of octylamine.

6.5
The Heck Reaction

Palladium-catalyzed olefin arylation reactions ("Heck coupling") have been successfully employed for the generation of C-C bonds in organic synthesis for decades [174–177]. Arylhalides and olefins are coupled by palladium catalysts (typically with phosphine co-ligands) in the presence of base, such as a trialkylamine.

Three different methodologies have been employed to extend the use of the Heck reaction to the synthesis of PPV, and they are presented in Scheme 45. Method A uses ethylene gas that is bubbled through a solution of previously activated Pd(0) catalyst, and an appropriately substituted dibromo- or diiodo-ben-

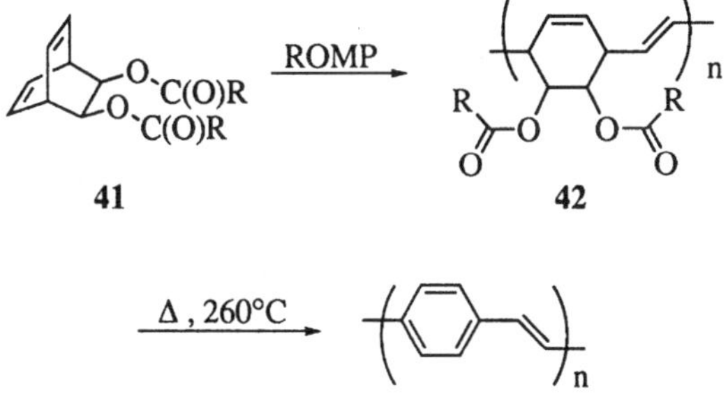

Scheme 44

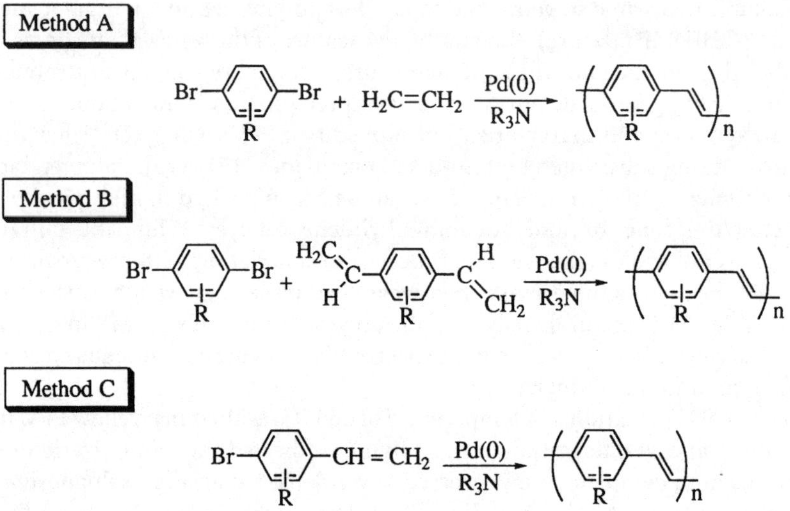

Method A

Br—[C₆H₂R]—Br + H₂C=CH₂ →[Pd(0), R₃N] —(—[C₆H₂R]—CH=CH—)—ₙ

Method B

Br—[C₆H₂R]—Br + H₂C=CH—[C₆H₂R]—CH=CH₂ →[Pd(0), R₃N] —(—[C₆H₂R]—CH=CH—)—ₙ

Method C

Br—[C₆H₂R]—CH=CH₂ →[Pd(0), R₃N] —(—[C₆H₂R]—CH=CH—)—ₙ

Scheme 45

Br–Ar–Br + H₂C=CH₂ →[Pd(0), R₃N] —(—Ar—CH=CH—)—ₙ

Ar = phenylene, 2-fluorophenylene, 2-nitrophenylene, 2-fluoro-5-methylphenylene, 2-phenylphenylene, biphenylene, 2,2'-dimethylbiphenylene

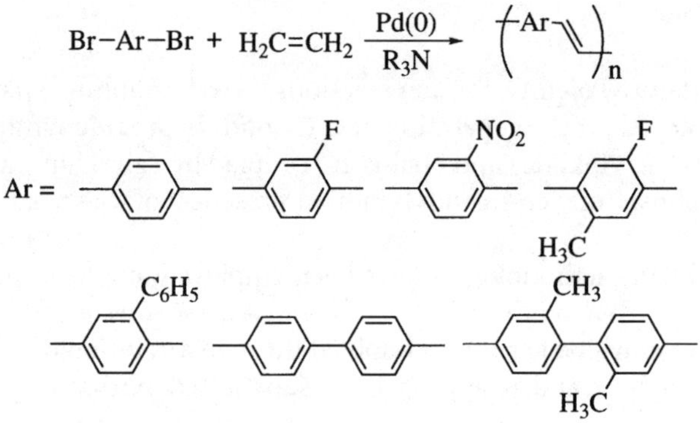

Scheme 46

zene. Perhaps the most widely employed approach, method B, utilizes both divinylbenzene and dibromobenzene as substrates that are coupled in the presence of a Pd(0) catalyst. Finally, PPV can also be synthesized by method C, where an A-B type difunctional monomer is self-coupled.

Substituted-PPVs and a number of PPV copolymers have been synthesized, illustrated in Scheme 46, using the Heck reaction shown above [178–180]. Of these polymers both the phenyl-substituted PPV and the biphenylene vinylenes are soluble in organic solvents while the methyl-, trifluoromethyl-, nitro- and fluoro-substituted analogs displayed poor solubility.

Thermotropic liquid crystalline PPV derivatives **43** were prepared by the coupling of dihalodialkoxybenzene and divinylbenzene in the presence of a palladium catalyst, as outlined in Scheme 47 [133]. Polarized light microscopy, as a function of temperature, showed evidence of a nematically ordered structure in the material. X-ray diffraction analysis of the pristine polymers showed them to be semi-crystalline in nature, although the crystallinity of the polymer changed dramatically upon heating above 100 °C.

The Heck reaction appears well-suited for the synthesis of PPV derivatives for nonlinear optical applications. In Scheme 48, a difunctional monomer **44**, con-

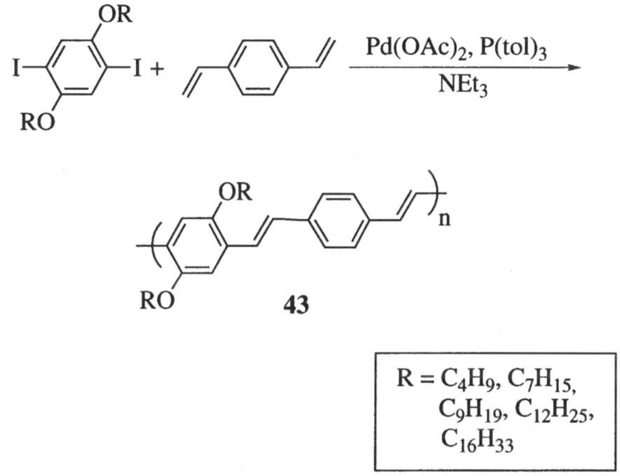

$R = C_4H_9, C_7H_{15},$
$C_9H_{19}, C_{12}H_{25},$
$C_{16}H_{33}$

Scheme 47

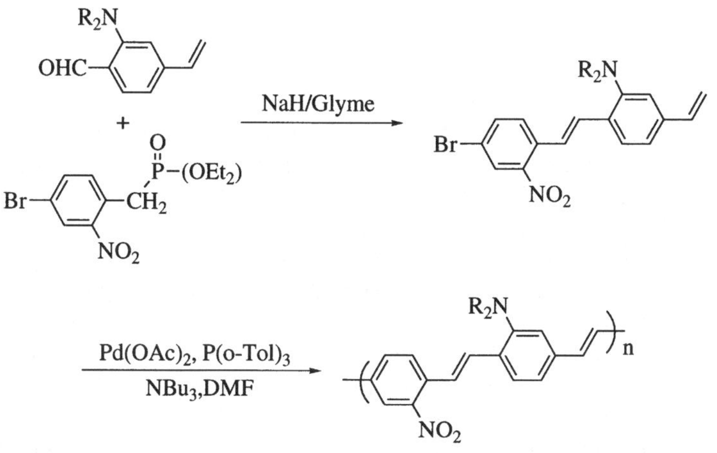

Scheme 48

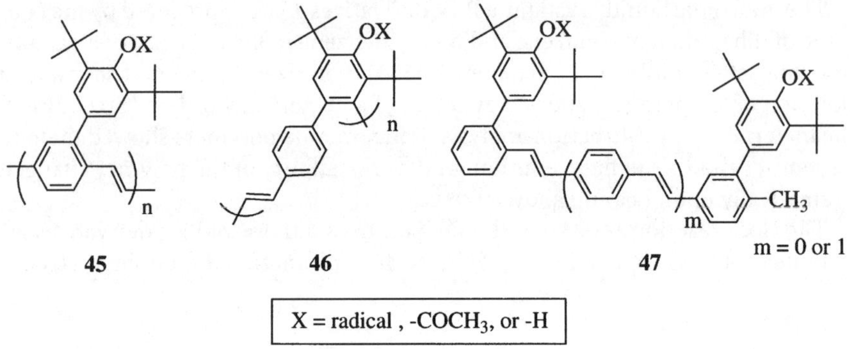

Scheme 49

X = radical , -COCH₃, or -H (shown as: $X = \text{radical}, -COCH_3, \text{or } -H$)

taining an active NLO chromophore, is first synthesized using traditional Horner-Emmons chemistry and polymerized using standard Heck coupling conditions.

Recently, the synthesis of polyradicals with a PPV backbone has been reported. Both 1,4- and 1,2-phenylenevinylenes, bearing phenoxy groups, were polymerized using the Heck reaction, and the structures of the PPV oligomers 45-47 are shown in Scheme 49 [181–183]. All of these materials were obtained as yellow powders, being soluble in common organic solvents. The acetylated phenols could be hydrolyzed with base and oxidized by PbO_2 to give radical species, as brownish/green powders. ESR spectra showed broad structure, suggesting that the spins were delocalized over two phenyl rings.

7
Polythiophenes

7.1
Introduction

Polythiophenes (PTs) have received a great deal of attention due to their electrical properties, environmental stability in doped and undoped states, non-linear optical properties, and highly reversible redox switching [1]. Thiophene possesses a rich synthetic flexibility, allowing for the use of several polymerization methods and the incorporation of various side chain functionalities. Thus, it is of no great surprise that PTs have become the most widely studied of all conjugated polyheterocycles [184].

The polymerization of thiophene, to yield intractable polythiophene **48**, was first carried out in a controlled manner in the early 1980s [185–188]. Even in such an unyielding form, this polymer displayed many promising optical and electronic properties. Unfortunately, its lack of processibility precluded further exploration of these attractive attributes. Since then, the synthesis of soluble pol-

Structure 48

$$\text{[structure 48: polythiophene repeat unit]}$$
48

ythiophene analogs has been the impetus for a tremendous amount of work, and has motivated a significant number of technological advances in this area. Most goals have been aimed at developing new types of polymerization techniques to attain improved polymer yields, greater synthetic utility, and enhanced physical and electronic properties.

As outlined earlier, three methods of polymerization have been established for the preparation of thiophenes, viz. electrochemical polymerization [189, 190], oxidative chemical polymerization using Lewis acid catalysts such as $FeCl_3$ [191, 192], and step-growth condensation polymerization using transition metal-catalyzed coupling reactions [1j].

7.2
Lewis Acid-Induced Polymerizations

The chemical oxidative polymerization of thiophene was initially reported using Lewis acid catalysts including $FeCl_3$, $MoCl_5$, and $RuCl_3$ [191, 192]. Such Lewis acids are readily available, and many are relatively inexpensive. Polymerization is quite facile using this methodology, performed by stirring monomer with excess oxidant. Many PTs, synthesized using $FeCl_3$, boast relatively high molecular weights, in some cases much higher than the same polymers obtained using the more delicate cross-coupling reactions. Thus, it is not surprising that $FeCl_3$-induced polymerization is now one of the most commonly used polymerization technique for the synthesis of polythiophenes.

However, this method is not without its drawbacks, which sometimes prove quite detrimental to polymer properties. Most notably is the occurrence of "coupling defects" along the π-backbone [193, 194]. The most conducting polythiophene backbone consists exclusively of α-site linkages in a highly planar arrangement. With unsubstituted β-sites present in the parent monomer, significant degrees of defective couplings occur at these positions during polymerization, greatly disrupting the polymer's effective conjugation length. Under certain conditions, polythiophenes produced via the Lewis acid route have also been found to contain residual impurities from the polymerization agent. For example, while the $Ce(SO_4)_2$ or $(NH_4)_2Ce(NO_3)_6$ oxidants have proven to rapidly polymerize 3,4-ethylenedioxythiophene, they also result in significant amounts of cerium-based residues in the final polymer [195]. Use of $Ce(SO_4)_2$ and H_2SO_4 (to help solubilize the oxidant) and long polymerization times yielded almost 45 wt% of cerium-based impurity.

7.3
Transition Metal-Mediated Coupling Polymerization

The chemistry of the various condensation polymerizations utilizing thiophene-based monomers is similar to that utilized in the synthesis of polyphenylenes, as discussed in Sect. 5. Coupling reactions including the Suzuki coupling [14], Stille coupling [12, 196], and Yamamoto reaction [1j] have been successfully employed for the synthesis of substituted PTs. (Detailed chemistry of these reactions was discussed earlier in this review, and so, the following section will concentrate on new developments in the synthesis of PTs.) Two additional transition metal-mediated methods have been used to produce PTs and deserve attention here. The first is an extension of the Ullmann coupling reaction [197, 198]. This polymerization, shown in Scheme 50, utilizes copper powder and dihalide monomer [199]. This coupling methodology is an attractive alternative for the polymerization of heterocycles bearing strongly electron-withdrawing groups as they generally tend to enhance the reaction.

The remaining polymerization route involves zero-valent nickel complexes and dihalide monomers. Variations of this route most often arise where different sources or regeneration methods of the active nickel species are utilized [82, 199, 200–204]. A typical example is shown below in Scheme 51 in which poly(3-phenylthiophene) **50** is synthesized from the parent 2,5-dichlorothiophene. As with the Ullmann reaction, polymerization appears to be most compatible with ring systems containing electron-withdrawing substituents.

While these polymerizations are generally more involved than chemical or electrochemical oxidative methods, well-defined monomers and highly selective transition metal catalysts yield polymers virtually free of β-coupling defects. Hence, the ability to obtain structurally superior materials has greatly fueled the search for optimized condensation reactions.

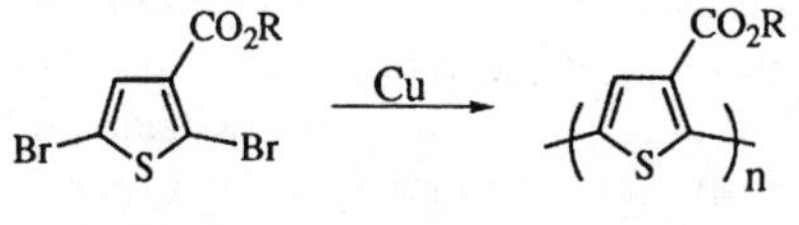

Scheme 50

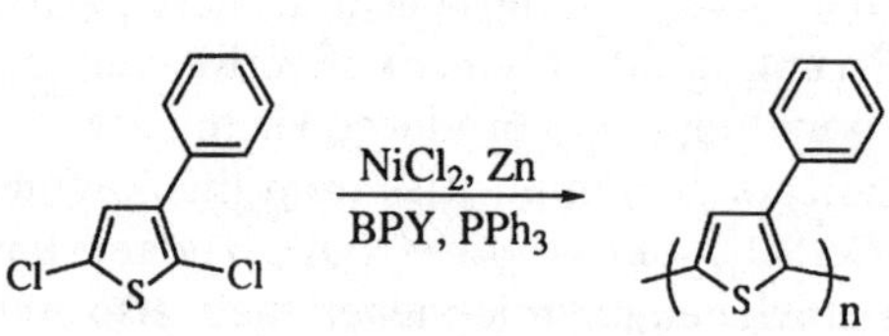

Scheme 51

7.4
Poly(alkylthiophenes)

With the realization of the attractive properties of polythiophene, a worldwide crusade in search of soluble analogs began. A soluble version was synthesized in 1986 that utilized repeat units bearing alkyl groups at the 3-position [205–208]. Since the report of these early poly(3-alkylthiophene)'s (P3AT), a large number of substituted polythiophenes have been synthesized incorporating solubilizing groups at the 3- or 3,4-positions.

Alkyl or aryl thiophenes **51** are typically synthesized "Kumada style" by the coupling of an alkyl or aryl Grignard reagent in a nickel(II)-catalyzed cross-coupling reaction with 3-bromothiophene as shown in Scheme 52 [209]. These substituted thiophenes can then be further funtionalized for polymerization utilizing standard organic methodologies (i.e. halogenation, stannylation, etc.).[210]

PATs containing alkyl groups with lengths ranging from methyl to cetyl groups have been synthesized. When the alkyl chain is longer than butyl, it affords polymers soluble in common organic solvents (i.e. tetrahydrofuran, chloroform, dichloromethane, trichloroethane and toluene). Synthesis of PTs functionalized with branched alkyls[211, 212], alkyl ethers,[213–216] esters [199, 203, 216–219], perfluorinated alkyls [220–223], chiral [205, 224], aryls [204, 225, 226], and self-doping sulfonate [227–230] groups have been reported.

While the incorporation of these solublizing groups at the 3-position has yielded PTs that have surmounted the intractability issue and also contain fewer β-defects, such substitution of the thiophene core necessarily results in the loss of ring symmetry. With these functionalized monomers, two different types of α-linkages now exist since the 2- and 5-positions are no longer equivalent [231, 232].

Head-to-tail (HT) linkages are formed upon coupling of the 2-position of one monomer to the 5-position of another. Analogously, head-to-head (HH) or tail-to-tail (TT) linkages are formed due to 2-, 2'- or 5-, 5'-couplings, respectively. The matter gets much more complicated when a triad is considered where, as shown in Scheme 53, four possible coupling schemes can occur. Side chain interactions in the non-HT-HT-coupled units induce twisting in the polythiophene backbone and disrupt the conjugation of the polymer's extended π-system.

Several attempts have been made to identify the regioregularity in poly(3-alkylthiophenes) synthesized using conventional methods, and to optimize the

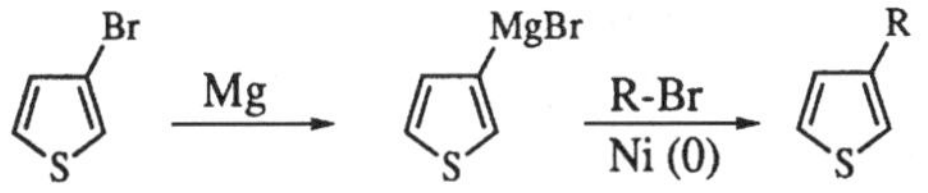

51

Scheme 52

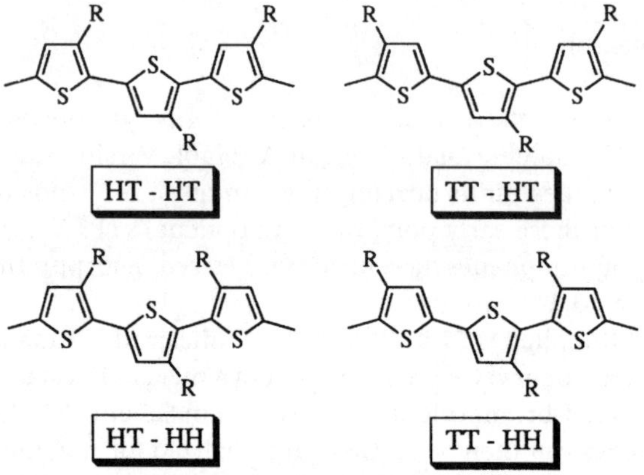

Scheme 53

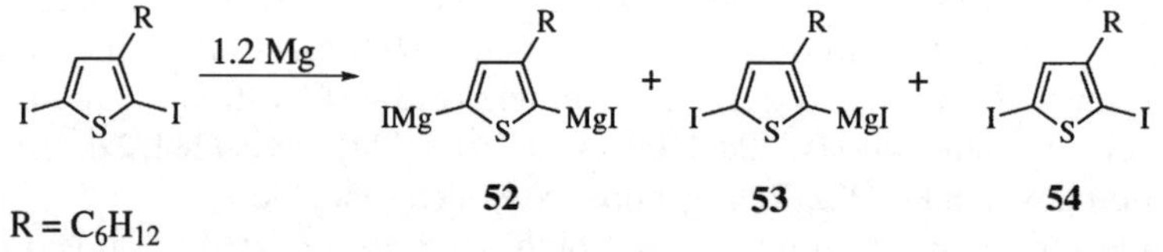

Scheme 54

reaction conditions to obtain the maximum composition of head-to-tail linkages. Regioregularity in poly(3-hexylthiophene), synthesized by the cross-coupling of the di-Grignard of 2,5-diiodothiophene, has been studied in detail [233]. When 2,5-diiodothiophene was treated with Mg metal (1:1 molar ratio) as indicated in Scheme 54, the reaction mixture contained approximately 43% di-Grignard **52**, 23% mono-Grignard **53** and 34% of unreacted diiodomonomer **54**. Polymerization of this reaction mixture afforded polymers of relatively low molecular weight (M_n ~1800 g/mol) highlighting the complications associated with controlled Grignard formation. When 1.2 equivalents of Mg per iodo group were used in a Yamamoto polymerization of 2,5-diiodo-3-(alkyl)thiophene, the polymer obtained consisted of 35% HT-HT coupling, whereas 0.6 equivalents of Mg per iodo group resulted in 58% HT-HT coupling [235]. However, the same polymer synthesized via a $FeCl_3$ polymerization using the unhalogenated monomer was found to contain more than 80% HT-HT coupling.

Early studies showed that polymers possessing some regiospecificity exhibit improved electrical and magnetic properties when compared to random polythiophenes [234]. In this instance, P3ATs **56** were synthesized by polymerization of 3,3'-dialkyl-2,2'-bithiophene **55** as shown in Scheme 55. Polymerization either electrochemically or via Lewis acid yielded predominantly HH-coupled polymers whose absorption maxima blue shifted ~90 nm for 3,3'-dimethyl-2,2'-

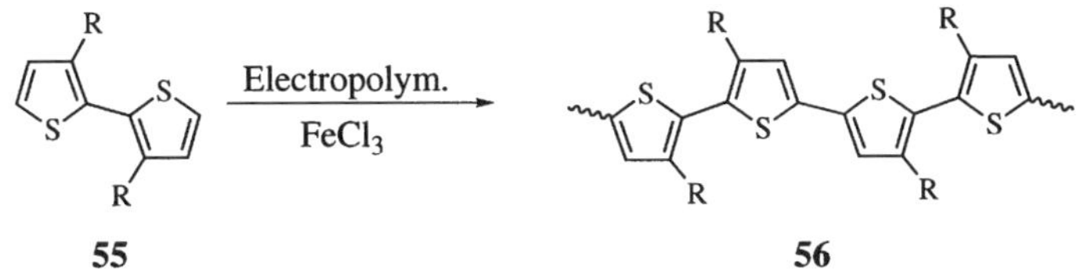

Scheme 55

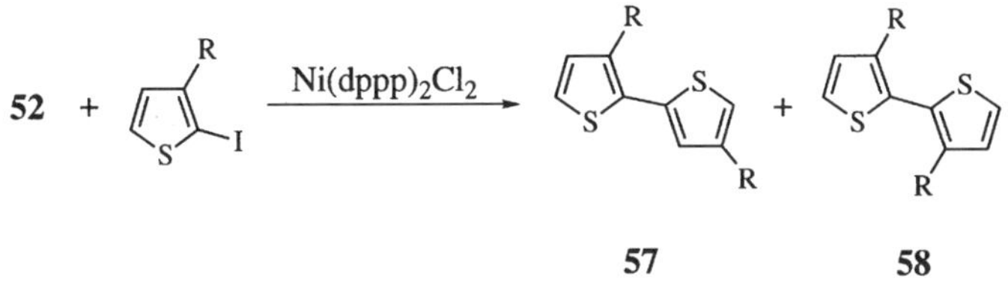

55 →(Electropolym. FeCl₃)→ 56

Scheme 56

52 + (thiophene-I with R) →(Ni(dppp)₂Cl₂)→ 57 + 58

bithiophene and ~110 nm for 3,3'-dihexyl-2,2'-bithiophene compared to that of normal poly(3-alkyl thiophenes).

7.5
Regioregular Polythiophenes

Several attempts have been made toward the synthesis of exclusively regioregular PTs using substituted bithiophene and terthiophene monomers. These multiring monomers can be synthesized using typical coupling chemistry as shown in Scheme 56. Polymerization of HH **57** or HT **58** coupled bithiophenes gives rise to some degree of regioregularity in the polymer, but the resulting materials still leave room for vast improvements.

Utilizing extremely controlled reaction conditions, the synthesis of regioselective poly(3-(4-octylphenyl) thiophene) **59** was reported using a FeCl₃-induced oxidative polymerization [236]. In this instance, highly regioregular **59** was synthesized by slow addition of FeCl₃ into the monomer solution. The resulting polymer was extracted with diethyl ether to remove irregular polymer with the resultant ether insoluble fraction consisting of highly regioregular polymer **59**. Comparison of the ^{1}H NMR spectra of the ether-soluble and -insoluble fractions indicated that ether-insoluble fractions had approximately 94% HT-HT coupling compared to the ether-soluble sample's 77% HT-HT constitution. However, the regiospecific nature of this method is quite sensitive to polymerization conditions. An increase in FeCl₃ concentration resulted in the loss of regioselectivity, indicating that formation of regioregular poly(3-phenyloctyl thiophene) is a function of oxidant concentration. Furthermore, only monomers containing a phenyloctyl side chain exhibited a high HT-HT-coupled composition while other regioselective polythiophene derivatives were not obtainable by slow FeCl₃ addition.

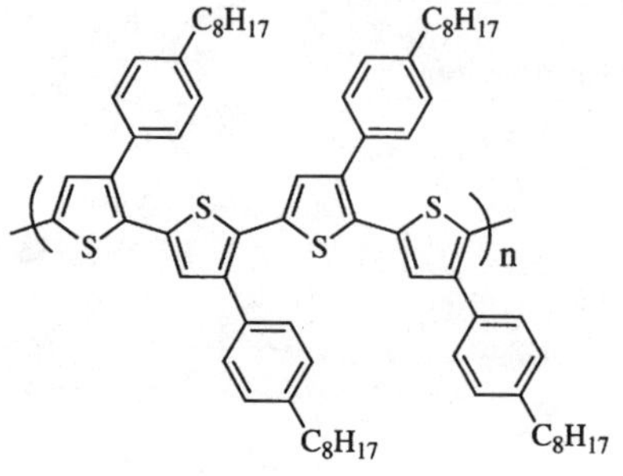
C8H17
C8H17
S
S
S
S
n
C8H17
C8H17

59

Structure 59

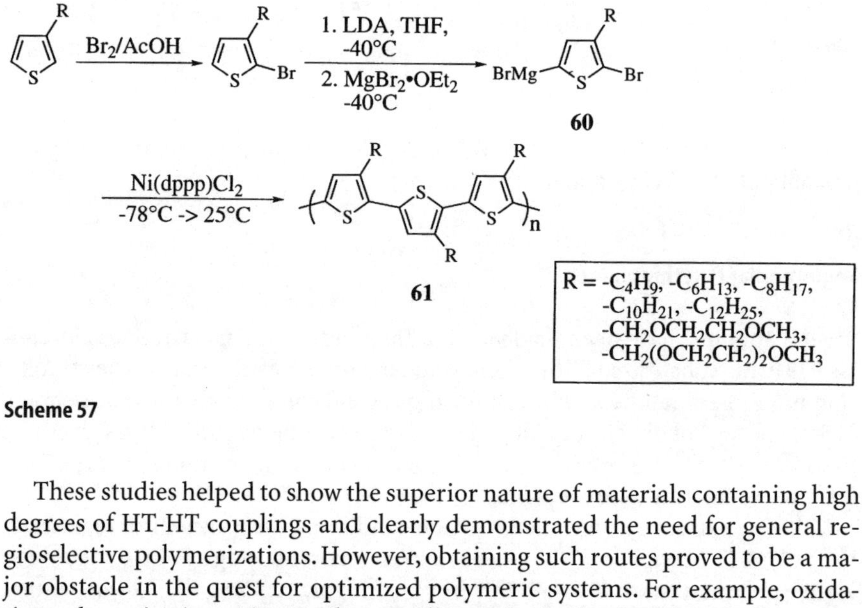
R
S
Br2/AcOH
R
S
Br
1. LDA, THF,
-40°C
2. MgBr2•OEt2
-40°C
R
BrMg
S
Br
60
Ni(dppp)Cl2
-78°C -> 25°C
R
S
S
R
S
n
R
61
R = -C4H9, -C6H13, -C8H17,
-C10H21, -C12H25,
-CH2OCH2CH2OCH3,
-CH2(OCH2CH2)2OCH3

Scheme 57

These studies helped to show the superior nature of materials containing high degrees of HT-HT couplings and clearly demonstrated the need for general regioselective polymerizations. However, obtaining such routes proved to be a major obstacle in the quest for optimized polymeric systems. For example, oxidative polymerization using $FeCl_3$ typically results in the formation of less than 80% of the desired HT-HT-linkages, and while electrochemical polymerization can show somewhat high selectivities under certain conditions, the corresponding polymers still contained significantly high levels of defective couplings [190, 230, 231, 237–239]. While the ratio of HT- to HH-linkages varied with reaction conditions and nature of the R-group, standard methodologies did not yield the sought after, exclusively HT-HT-coupled polymer.

Two different avenues have been pursued leading toward the synthesis of highly regioregular polythiophenes (>98% HT-HT). The first regioregular P3ATs **61** synthesized were comprised of nearly 100% HT-HT couplings using the Grignard reagent of **60** [216, 238, 240]. This route, shown in Scheme 57, has become known as the McCullough method, and utilizes exclusive bromination

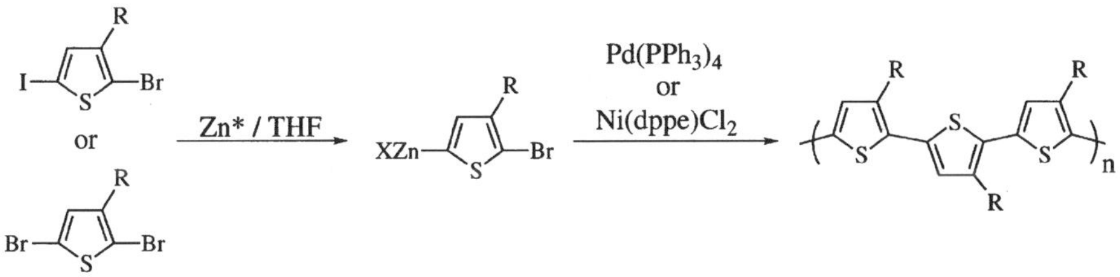

Scheme 58

at the 2-position of the 3-alkylthiophene precursor, selective lithiation at the 5-position with LDA followed by trapping with $MgBr_2 \cdot OEt_2$, and polymerization via a nickel catalyst. The stability of the A-B difunctional monomer against scrambling allows for precise control of coupling.

In the second method, activated zinc is employed for the synthesis of regioregular P3ATs [241–244]. This procedure is illustrated in Scheme 58 and revolves around the selectivity of Reike zinc addition followed by Ni(0)- or Pd(0)-catalyzed cross-coupling polymerization. Catalyst selection has been shown to play a major role in this coupling scheme. For example, $Pd(PPh_3)_4$ was determined to be the most effective catalyst for the coupling of iodozinc compounds, whereas $Ni(dppe)Cl_2$ proved superior in conjunction with bromozinc monomers.

Both ^1H and ^{13}C NMR spectroscopies are perhaps the best tools for the detection of HH- and HT-couplings. The ^1H NMR spectra of P3ATs, possessing purely regiorandom constitutions, show the presence of four peaks in the aromatic region [231, 232], whereas P3ATs, containing exclusively HT-linkages, display a single peak at approximately 7.0 ppm [240]. Accordingly, the ^{13}C NMR spectra of P3ATs formed solely from HT-couplings contain only four aromatic signals, while polythiophene synthesized using $FeCl_3$ shows the presence of more than 16 signals attributable to the four possible coupling types.

Not surprisingly, the electrical and optical properties of regioregular PTs differ greatly from their regiorandom counterparts [240, 241]. For example, the optical band gap of regioregular poly(3-butylthiophene) is 1.7 eV, whereas that of random poly(3-butylthiophene) is 2.1 eV. The electrical conductivities of I_2-doped films of regioregular poly(3-butylthiophene) have been reported as high as 1350 S cm^{-1} while that of random poly(3-butylthiophene) exhibited a maximum of 5 S cm^{-1}. The λ_{max} of the absorbance and emission spectra both exhibit red shifts of 20 nm in the case of regioregular poly(3-butylthiophene), indicating a greater effective conjugation length in the regioregular systems.

7.6
Poly(3,4-ethylenedioxythiophene) (PEDOT) and Derivatives

In order to increase the electron-rich character of monomer and polymer, leading to reduced oxidation potentials and lower band gaps, researchers at Bayer AG have developed PEDOT [245–248]. By appending the electron donating di-

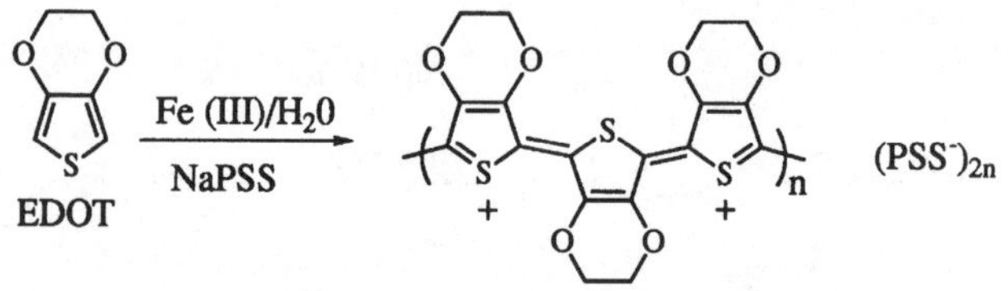

Scheme 59

oxyethane bridge across the 3,4 positions, sufficient HOMO raising is effected and polymerization is forced to occur through the open 2,5-positions yielding a linear, highly conjugated polymer. Prepared as an aqueous dispersion with poly(styrene sulfonate) dopant as shown in Scheme 59, PEDOT-PSS has become the most commercially successful of the conducting polymers. As a relatively low band gap (E_g=1.6 eV) polymer, the conducting form of PEDOT is quite transmissive to visible light and highly conductive (200 S cm^{-1}), allowing it to be used as an anti-stat coating in the photographic film industry [249–251].

A number of new polymerizable derivatives based on substituted EDOT, or molecules containing EDOT as the polymerizable unit in a multi-ring monomer, have been prepared by our group, and others. In this manner, organic soluble PEDOTs [252], water-soluble PEDOTs [253], high contrast electrochromic PEDOTs [254–256], and multi-color, variable band gap polymers have been developed [257–262]. Due to the ease of synthetic derivatization, polymerizability by both chemical and electrochemical methods, high stability of the oxidized conducting form, facility for redox switching, and ability to control optoelectronic properties in general, the PEDOT-based polymers provide significant opportunity for the future.

8
Polypyrroles

Comparable to thiophene, pyrrole is a five-membered heterocycle, yet the ring nitrogen results in a molecule with distinctly different behavior and a far greater tendency to polymerize oxidatively. The first report of the synthesis of polypyrrole (PPy) **62** that alluded to its electrically conductive nature was published in 1968 [263]. This early material was obtained via electrochemical polymerization and was carried out in 0.1 N sulfuric acid to produce a black film. Since then, a number of improvements, which have resulted from in-depth solvent and electrolyte studies, have made the electrochemical synthesis of PPy the most widely employed method [264–266]. The properties of electrosynthesized PPy are quite sensitive to the electrochemical environment in which it is obtained. The use of various electrolytes yield materials with pronounced differences in conductivity, film morphology, and overall performance [267–270]. Furthermore, the water solubility of pyrrole allows aqueous electrochemistry [271], which is of prime importance for biological applications [272].

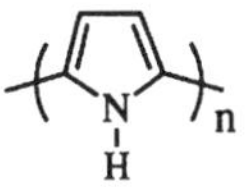

Structure 62

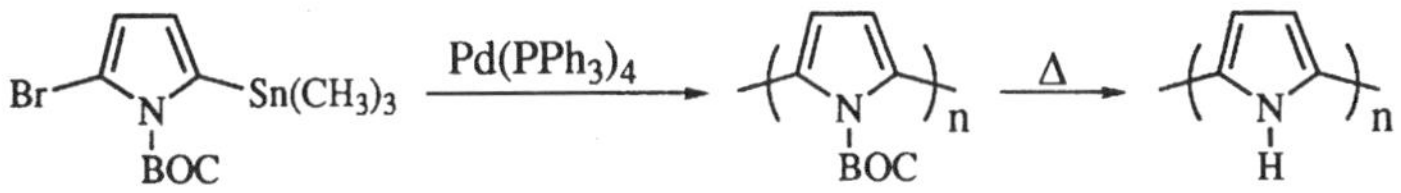

62

Scheme 60

63

PPy is also obtained by the treatment of pyrrole with chemical oxidants, often in the presence of charge-compensating dopant anions [273, 274]. Relative to PT, only a small body of research has addressed the chemical polymerization of pyrrole [275–277]. However, the more readily oxidized pyrrole moiety allows for the use of less stringent conditions than are required for PT synthesis [276, 278]. (This same characteristic also results in PPy being prone to degradative overoxidation [279].) Substitution of the pyrrole nitrogen affords soluble PPy derivatives bearing a variety of functionalities [280, 281]. Unfortunately, derivatization in the N-position creates detrimental steric interactions with adjacent rings, resulting in a substantial decrease in conducting properties [267].

While electrochemical syntheses and, to a lesser extent, chemical oxidant routes have provided many interesting materials, direct chemical polymerization to well-characterized polymers has been more limited. In an analogous manner to thiophene, efforts have been made to synthesize 3-substituted pyrrole monomers [273, 274, 282–286]. The syntheses in this case are somewhat more tedious, requiring initial protection of the pyrrole nitrogen (with groups such as tosylates or *tert*-butoxy carbonyl, BOC) and subsequent removal of the protecting group after functionalization. The ease of oxidation of pyrrole can make many of these synthetic manipulations difficult as reaction with impurities prior to, and during purification can result in a dark tar known as "pyrrole black". However, it has been shown that 3-substituted PPys display significantly improved properties over their N-substituted analogs.

PPys synthesized by both oxidative routes are also subject to coupling defects, which drastically reduce sought after properties. To circumvent this problem, transition metal-mediated polymerizations have been explored. Once again, the synthetic inflexibility of the pyrrole moiety has proven to be a formidable obstacle in obtaining such materials. The Stille coupling scheme [12], shown in Scheme 60, has been used to prepare a BOC-substituted PPy **63** with the protecting group subsequently removed by thermolytic treatment to yield unsubstitut-

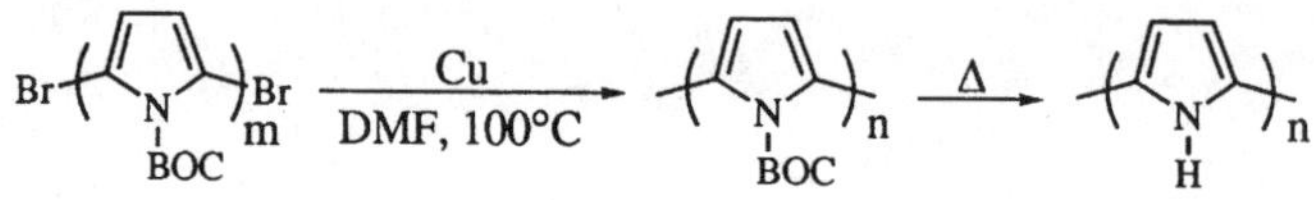

Scheme 61

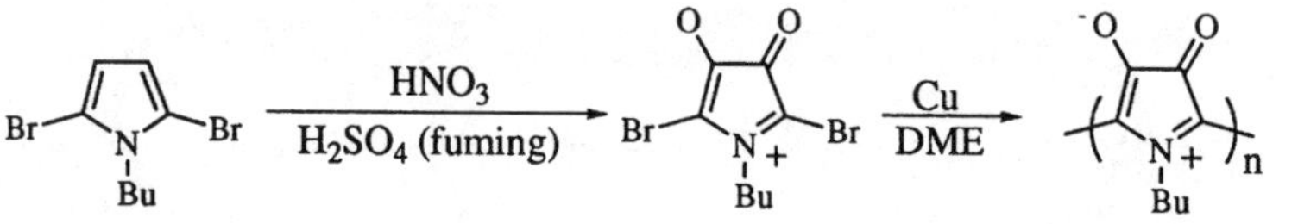

Scheme 62

ed-Ppy [287, 288]. Materials obtained in this manner were of modest molecular weight (ca. 16 repeat units), but they have been shown to be structurally well-defined containing exclusively 2,5-linkages.

The polymerization of a protected pyrrole has been reported based upon the Ullman coupling reaction [215, 216, 289]. As illustrated in Scheme 61, the 2,5-dibromo-N-BOC-protected monomer was treated with CuCl followed by subsequent thermolysis to yield materials similar to those prepared in Scheme 60. Fractionation via HPLC made it possible to separate and characterize pyrrole oligomers up to $n=20$.

This coupling methodology was subsequently utilized to prepare zwitterionic-PPy **64** that possessed a remarkably low solution band gap of 1.1 eV [272]. Their strategy, depicted in Scheme 62, included a copper-bronze-promoted polymerization to afford polymer **64**, exhibiting an M_n of ca. 5000 g/mol relative to polystyrene. While this material was reported to be an intrinsic semiconductor, it showed interesting pH-dependent changes in the electronic absorption spectra.

It is evident from the properties accessible that PPys may prove to be very useful materials in the future. To date, the synthetic difficulties encountered have thwarted progress and made processible and well-characterized, high molecular weight systems largely untenable. Future breakthroughs are required in this area to overcome these obstacles.

9
Polyaniline

Polyaniline (PANI) is perhaps the oldest of the conducting polymers. References dating back to 1862 can be found describing a material known as "aniline black" [291]. A few reports surfaced on PANI in the 1960s [292–294], most notably a study investigating the effects of acids on PANIs conductivity [295], but it was not until the 1980s that its electrically conducting properties became fully appreciated [296]. PANI has become an extremely well-studied CP owing to its low cost and remarkable stability under a variety of conditions [297, 298]. This has resulted in PANI being the conjugated polymer of choice for many technological applications and has led to its development as a commercial product. There are several forms of PANI, and these are shown in Fig. 3. Green and Woodhead first documented and named these various oxidation states, and their nomenclature continues to be used today [299, 300].

PANI is unique in that its most oxidized state, the pernigraniline form (which can be accessed reversibly), is not conducting. In fact, it is the intermediately oxidized emeraldine base that exhibits the highest electrical conductivity. "Protonic Acid Doping" is the most general means by which to obtain this partially protonated form of PANI [301]. Exposure of the emeraldine salt to alkali solutions reverses this process and brings a return to the insulating state.

Both electrochemical and chemical oxidative routes are most often utilized for the synthesis of PANI. In an interesting departure from the oxidative route, poly(phenylene amine imine) was prepared via a conventional condensation polymerization, as illustrated in Scheme 63 [302, 303]. Comparison of this structurally well-characterized polymer with oxidatively prepared PANI allowed confirmation of the PANI structure. However, the structure of PANI produced by electrochemical means is less understood.

Electropolymerization in acidic media affords free-standing films that are believed to contain varying degrees of cross-linking [267, 292, 304]. The miscibility of aniline with water allows for a variety of aqueous oxidants, such as ammonium peroxydisulfate, to be used [305]. Chemical polymerization of aniline can also be performed in chloroform through the use of tetrabutyl ammonium periodate [306]. Accordingly, a number of alkyl [301] and alkoxy-substituted [307] aniline derivatives have been chemically polymerized. Unfortunately, functionalization of the aniline nucleus often leads to a decrease in performance in the resulting polymers [308, 309].

As with PPy, synthetic limitations have greatly limited the development of new PANI derivatives. The fruition of new, non-oxidative routes leading to PANI would, unquestionably, create an entirely new realm of possibilities. Furthermore, the ability to engineer enhanced functionality into the PANI nucleus without concomitant loss in polymer properties also represents a monumental task that, when accomplished, could help propel PANI to an even higher level of importance.

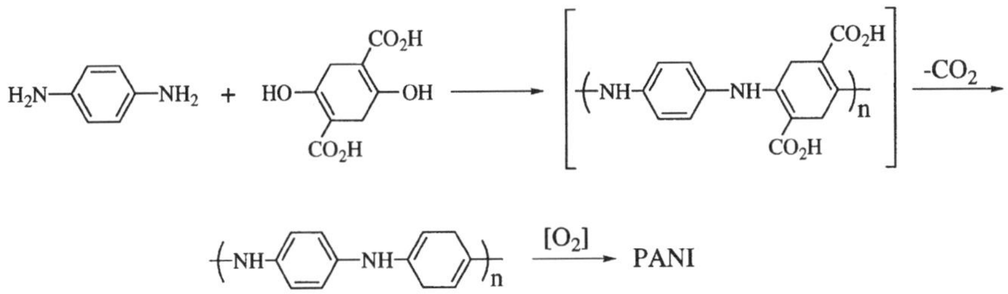

Scheme 63

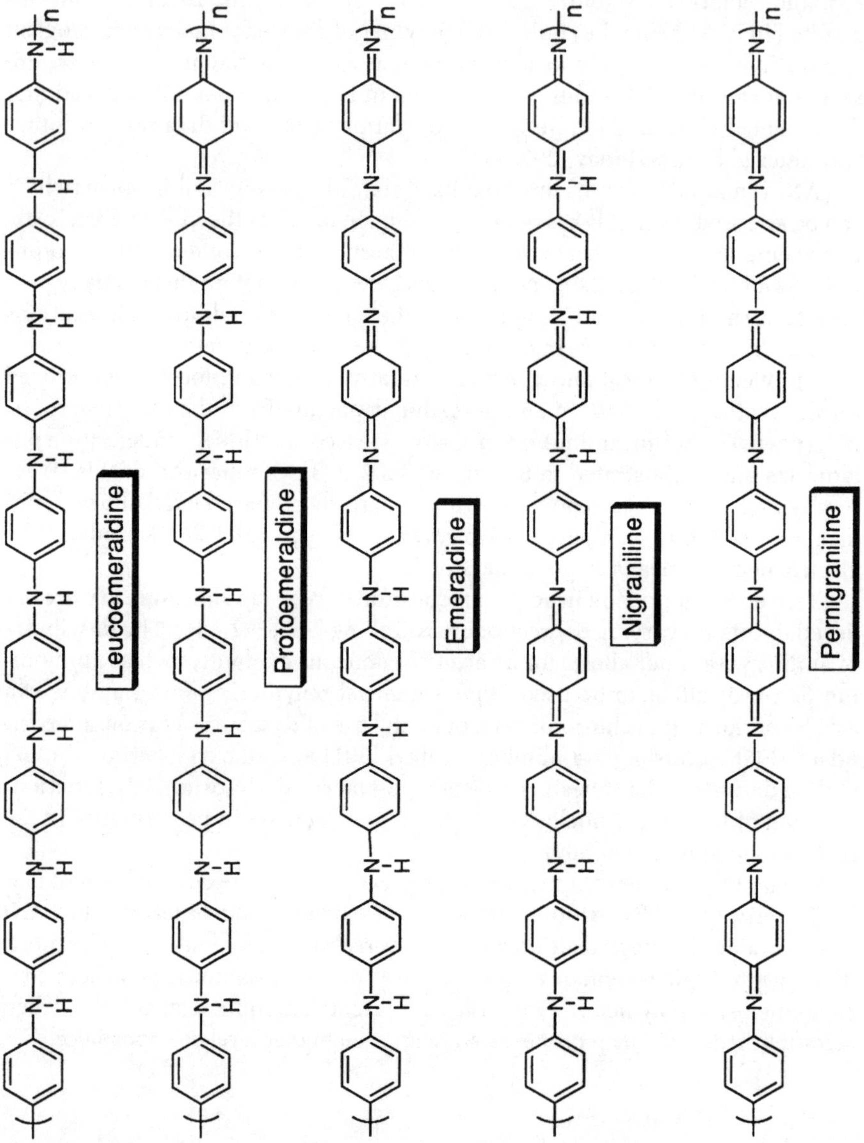

Fig. 3. Various forms of polyaniline (PANI)

10
New Vistas

This last section contains a broad collection of examples of CPs in order to introduce the reader to some of the possibilities that are afforded by a conjugated architecture. While some polymers have been known for many years (and are presented here because they do not belong to one particular class of CPs), others are quite new ventures, representing the vanguards of the field. Unfortunately, to mention each worthy new development would require far more space than is afforded in this already lengthy review, and so, only a few selected highlights and "hot areas" can be included.

10.1
Copolymers

The synthesis of conjugated materials from the polymerization of two monomer types to yield conducting copolymers has received tremendous attention over the years [310]. This material class is particularly well-suited for tuning the optoelectronic properties of conjugated polymers through the use of components possessing a broad range of electron densities. When transition metal-mediated routes are selected, copolymers can be synthesized with well-defined structures affording a variety of properties not available with their parent homopolymers.

The arylene-vinylene systems include a vast array of hybrid systems owing to their synthesis from classical organic means [311–314]. However, new systems continue to surface that are not viable other than from non-transition metal-mediated pathways. Just a few of the many examples obtained from the various synthetic routes mentioned previously in the PPV section are depicted below in Scheme 64.

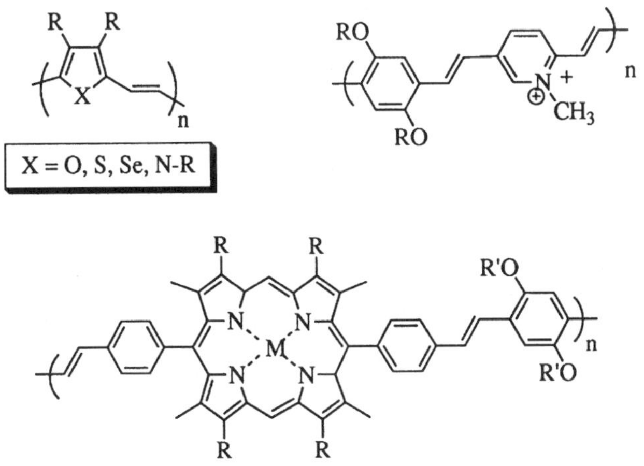

Scheme 64

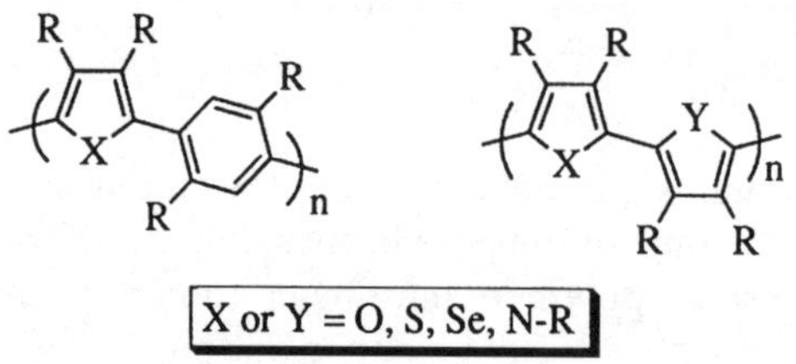

Scheme 65

X or Y = O, S, Se, N-R

A plethora of copolymers have been synthesized using many of the other cross-coupling reactions used in earlier sections [310]. Some of the more prominent structure types are shown above in Scheme 65 [315–320]. The possibilities for new π-conjugated copolymers seem endless as a huge combination of monomers can be coupled, being restricted only by polymerization-compatible functional groups.

10.2
Conjugated Oligomers

The interest in conjugated oligomers of discrete length as both model compounds and device components continues to grow [321]. Reports in the literature can be found documenting the synthesis of oligoanilines [302], -phenylenes [322], -pyrroles [323–326], and -thiophenes [327–331]. Oligothiophenes are undoubtedly the flagship example when addressing conjugated oligomers. The methods of preparation of these discreet molecules are very similar to those routes discussed in the previous sections with the transition metal-mediated couplings being the most widely employed. As one would expect, the properties of an oligomer is greatly a function of length. Oligothiophenes of up to six rings have been shown to polymerize with longer species being less reactive [332]. Likewise, it has also been shown that oligothiophenes containing 11 or more rings exhibit electronic properties in the range of PT [333, 334].

10.3
Cyclization of Prepolymers

As utilized in the synthesis of PPV-based polymers, soluble precursor routes have been developed for the synthesis of various heterocyclic π-conjugated polymers. The two most widely employed of these methods for heterocycle formation, shown in Scheme 66, center around ring closure of pre-polymers containing diacetylene **65** or 1,4-diketone units **66** [335–339]. The synthesis of heterocyclic structures from 1,4-diketones has been a known transformation in organic chemistry for decades. While once mainly used for monomer preparation through various cyclizations, it is now being employed to make heterocycle-containing polymers and copolymers [340–342].

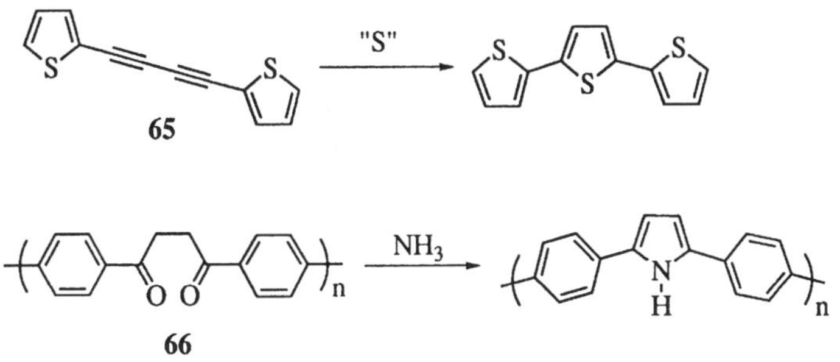

Scheme 66

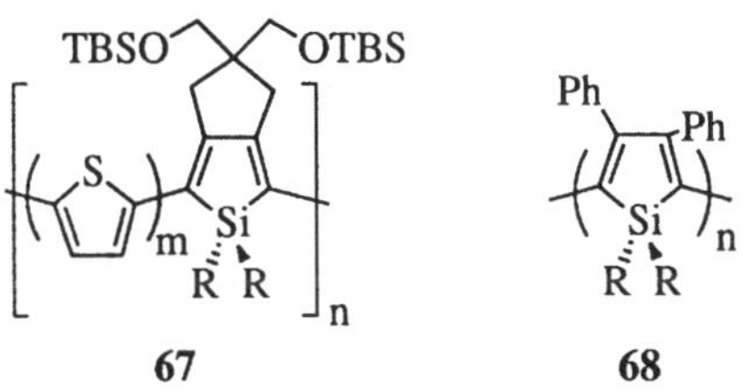

65, **66**

Scheme 66

67, **68**

Scheme 67

10.4 Silole Systems

Interest in silole-containing conjugated materials continues in hopes of exploiting the unique electronic properties of the silole ring (an unusually low-lying LUMO) for the development of novel systems. Recent advances regarding the synthesis of silole-containing polymers **67** and copolymers **68** have made these materials more accessible [343–346]. While inital work centered around the reductive cyclization/silylation of diynes [347], the synthesis of halogenated and stannylated silole monomers for use in typical transition metal-catalyzed cross-coupling reactions provides building blocks for a broad family of materials. Accordingly, a variety of silole-thiophene co-oligomers and copolymers have been synthesized, some of which are shown above in Scheme 67.

10.5 Transition Metal-Containing CPs

Metal-containing polymers, in which the metal is coordinated to the polymer chain, are interesting from the standpoint that the conjugated backbone can help stabilize redox activity, affording a continuum of accessible states. The very nature of these materials suggests an array of potential uses (e.g., electro-catalysis, electrochromic displays, and molecular recognition). There are many examples of π-conjugated polymers in the literature that possess pendant metal centers, attached to the backbone by insulating tethers [348]. For the most part metal/polymer interactions in these systems are inherently weak. To reap the greatest rewards of a transition metal/conjugated polymer hybrid, the ideal structure would have the metal centers directly affixed to, and in direct electronic communication with, the polymer backbone. As yet, there are still only a few reported examples of conducting and electroactive polymers where metal centers are in conjugation with the polymer's π-system. All of these systems possess metal centers coordinated to bidentate, nitrogen-containing, heterocyclic units (2,2'-bithiazole, 2,2'-bipyridyl, or Schiff base) incorporated into the polymer backbone. While the number of examples are few, these preliminary efforts have already shown the broad flexibility that metal coordination can impart to traditional organic systems.

Such design concepts were first reported utilizing 2,2'-bithiazole and 2,2'-bipyridine units, respectively, as postpolymerization metal coordination sites [349, 350]. Subsequently, a poly(p-phenylenevinylene)-based polymer **69** containing ionic ruthenium centers bound to bipyridyl (BPY) units incorporated into the polymer backbone was reported. This system, depicted in Scheme 68, exhibits enhanced photoconductivity relative to the parent organic polymer [351] (Yu).

Recently, metal/polymer complexation has been utilized as a means to polyrotaxane formation via a Sauvage-type [352] template effect [353, 354]. An elegant study showing the sensitivity of a bipyridine-containing, pseudo-poly(phenylenevinylene) system capable of complexing various metal ions followed [355]. In this work, conformational changes of the polymer, which are associated with the coordination of the metal ions, afforded a system that can toggle between its conjugated and nonconjugated forms.

Our group has recently become interested in these novel systems from an electrochromic and sensor standpoint. Our polymers are centered around bis(salicylidene)thienyl cores that can undergo site-directed electro-polymerization to

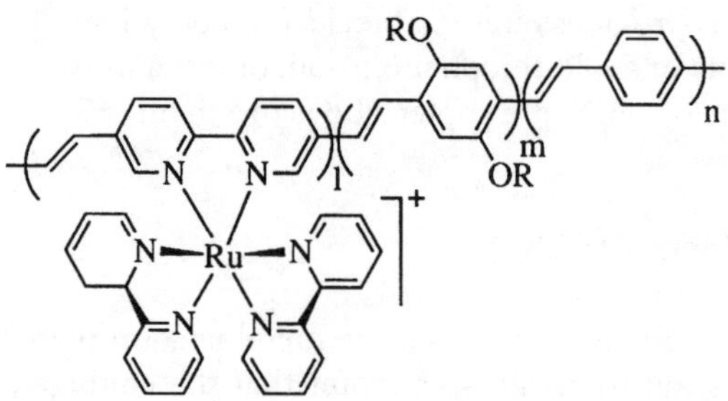

69

Scheme 68

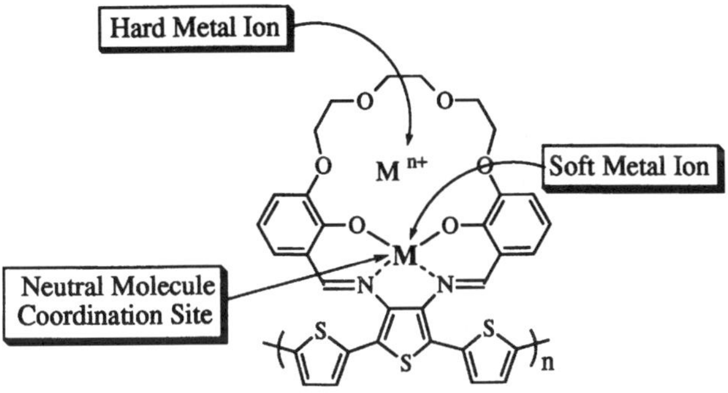

Scheme 69

yield phenylene- or thienylene-linked polymers [356, 357]. Recently, we have developed crown ether-containing analogues **70**, shown in Scheme 69, affording polymers capable of coordinating/sensing hard and soft metal ions, as well as neutral organic molecules [358].

11
Perspectives

As can be evidenced from this article, a tremendous amount of effort has served to provide a number of quantum leaps in the field of π-conjugated polymers. In fact, such a retrospective look at the synthetic advancements that have occurred in the past two decades might leave one with a better appreciation of the scientific method. Indeed, we have come a long way since the first reports of those mysterious and intractable black powders. The synthesis of well-charcterized soluble/processable materials with controlled properties has become routine. Application technology has also grown in leaps and bounds affording tangible evidence as to the potential that lies in this field. It is truly exciting to think of what the next 20 years may bring.

Acknowledgments. The authors wish to thank Dr. Jayesh Dharia for his efforts in the early stages of this manuscript. Funding was provided by the National Science Foundation (CHE 9629854) and the Air Force Office of Scientific Research (F49620-96-1-0067). Their financial assistance is greatly appreciated.

12
References

1. (a) Kovacic P, Jones MB (1987) Chem Rev 87: 357; (b) Patil AO, Heeger AJ, Wudl F (1988) Chem Rev 88: 183; (c) Reynolds JR (1988) Chemtech 18: 440; (d) Billingham NC, Calvert PD (1989) Adv Poly Sci 90: 1; (e) Kanatzidis MG (1990) Chem Eng News 68:

(49)36; (f) MacDiarmid AG, Epstein AJ (1991) Makromol Chem Makromol Symp 51: 11; (g) Baughman RH (1991) Makromol Chem Makromol Symp 51: 193; (h) Syed AA, Dinesan MK (1991) Talanta 38: 815; (i) Reynolds JR, Pomeranz M (1991) Electroresponsive Molecular and Polymeric Systems 2: 187; (j) Yamamoto T (1992) Prog Polym Sci 17: 1153; (k) Roncali J (1992) Chem Rev 92: 711; (l) Schlüter AD, Wegner G (1993) Acta Polym 44: 59; (m) Naarmann HJ (1993) Polym Sci: Polym Symp 75: 53; (n) Lux F (1994) Polymer 35: 2915; (o) Tour JM (1994) Adv Mater 6: 190; (p) Toshima N, Hara S (1995) Prog Polym Sci 20: 155; (q) Feast W J, Tsibouklis J, Pouwer KL, Groenendaal L, Meijer EW (1996) Polymer 37: 5017; (r) Pelter A, Jenkins I, Jones DE (1997) Tetrahedron 53: 10357; (s) Roncali J (1997) Chem Rev 97: 173
2. (a) Miller JS (1993) Adv Mater 5: 587; (b) Miller JS (1993) Adv Mater 5: 671
3. Peierls RE (1955) In Quantum Theory of Solids. Oxford University Press, Oxford
4. Shimamura K, Karasz FE, Hirsch JA, Chien JCW (1981) Makromol Chem Rapid Commun 2: 473
5. Hill MG, Penneau JF, Zinger B, Mann KR, Miller LL (1992) Chem Mater 4: 1106
6. Bäuerle P, Segelbacher U, Maier A, Mehring M (1993) J Am Chem Soc 115: 10217
7. Furukawa Y (1996) J Phys Chem 100: 15644
8. Grubbs RH (1982) In: Wilkinson G, Stone FGA, Abel EW (eds) Comprehensive Organometallic Chemistry. Pergamon, New York, p. 502
9. Yamamoto A (1986) In: Organotransition Metal Chemistry. Wiley, New York
10. Dragutan V, Balaban AT, Dimonie M (1986) In: Olefin Metathesis and Ring Opening Polymerization of Cycloolefins. Wiley, New York
11. Collman JP, Hegedus LS, Norton JR, Finke RG (1987) In: Principles and Applications of Organotransition Metal Chemistry. University Science Books, Mill Valley
12. Stille JK (1986) Angew Chem, Intl Ed Engl 25: 508
13. Yamamoto T, Hayashi Y, Yamamoto Y (1978) Bull Chem Soc Japan 51: 2091
14. Miyaura N, Suzuki A (1995) Chem Rev 95: 2457
15. Shirakawa H, Louis EJ, MacDiarmid AG, Chiang CK, Heeger AJ (1977) J Chem Soc, Chem Commun: 578
16. Chiang CK, Fincher CR Jr, Park YW, Heeger AJ, Shirakawa H, Louis EJ, Gau SC, Mac Diarmid AG (1977) Phys Rev Lett 39: 1098
17. MacDiarmind AG, Heeger AJ (1979) Synth Met 1: 101
18. Wegner G (1981) Angew Chem, Int Ed Engl 20: 361
19. Wynne KJ, Street GB (1982) I&EC Product Research and Development 21: 23
20. Baughman RH (1984) In: Vandenberg EJ (ed) Contemporary Topics in Polymer Science Vol 5. Plenum Publishing
21. Greene RL, Street GB (1984) Science 226: 651
22. Chien JCW (1984) In: Polyacetylene: Chemistry, Physics and Materials Science. Academic Press, Orlando FL
23. Krivoshei IV, Skorobogatov VM (1991) In Polyacetylene and Polyarylenes, Polymer Monographs Vol 10. Gordon and Breach
24. Gibson HW (1986) In: Skotheim TA (ed) Handbook of Conducting Polymers Vol 1. Marcel Dekker Inc, New York, pp. 405-439
25. Chien JCW, Wnek GE, Karasz FE, Hirsch JA (1981) Macromolecules 14: 479
26. Masuda T, Higashimura T (1984) Accts Chem Res 17: 51
27. Jin SH, Choi SJ, AhnW, Cho HN, Choi SK (1993) Macromolecules 26: 1487
28. Choi SJ, Jin SH, Park JW, Cho HN, Choi SK (1994) Macromolecules 27: 309
29. Simionescu CI, Dumitrescu S, Grigoras M, Percec V (1979) J Polym Sci, Polym Lett Ed 17: 287
30. Gibson HW, Bailey FC, Epstein AJ, Rommelman H, Kaplan S, Harbour J, Yang XQ, Tanner DB, Pochan JM (1983) J Am Chem Soc 105: 4417
31. Gal YS, Choi SK (1988) J Polym Sci, Polym Lett C 26: 115
32. Yakhimovich RI, Shilov EA, Dvorko GF (1966) Dokl Akad Nauk SSSR 166: 388
33. Kang KL, Kim SH, Cho HN, Choi KY, Choi SK (1993) Macromolecules 26: 4539

34. Kim SH, Choi SJ, Park JW, Cho HN, Choi SK (1994) Macromolecules 27: 2339
35. Subramanyam S, Blumstein A (1991) Makromol Chem Rapid Commun 12: 23
36. Subramanyam S, Blumstein A (1991) Macromolecules 24: 2668
37. Subramanyam S, Li KP, Blumstein A (1992) Macromolecules 25: 2065
38. Subramanyam S, Blumstein A (1992) Macromolecules 25: 4058
39. Subramanyam S, Chetan MS, Blumstein A (1993) Macromolecules 26: 3212
40. Edwards JH, Feast WJ (1980) Polymer 21: 595
41. Edwards JH, Feast WJ, Buff DC (1984) Polymer 25: 395
42. Swager TM, Dougherty DA, Grubbs RH (1988) J Am Chem Soc 110: 2973
43. Swager TM, Grubbs RH (1989) J Am Chem Soc 111: 4413
44. Ginsburg EJ, Gorman CB, Marder SR, Grubbs RH (1989) J Am Chem Soc 111: 7621
45. Klavetter FL, Grubbs RH (1989) J Am Chem Soc 111: 7807
46. Gorman CB, Ginsburg EJ, Sailor MJ, Moore JC, Jozefiak TH, Lewis NS, Grubbs RH, Marder SR, Perry JW (1991) Synth Met 41: 1033
47. Sailor MJ, Ginsburg EJ, Gorman CB, Kumar A, Grubbs RH, Lewis NS (1990) Science 249: 1146
48. Moore JS, Gorman CB, Grubbs RH (1991) J Am Chem Soc 113: 1704
49. Jozefiak TH, Ginsburg EJ, Gorman CB, Grubbs RH, Lewis NS (1992) J Am Chem Soc 115: 4705
50. Noren GK, Stille JK (1971) Makromol Rev 5: 385
51. Ried W, Freitag D (1966) Naturwiss 53: 305
52. Stille JK, Gilliams Y (1971) Macromolecules 4: 515
53. Stille JK (1972) Makromol Chem 154: 49
54. Krigbaum WR, Krause KJ (1978) J Polym Sci, Polym Chem Ed 16: 3151
55. Dineen JM, Howell EE, Volpe AA (1982) Polym Sci, Polym Chem Ed 23: 282
56. Claesson S, Gehm R, Kern W (1949) Makromol Chem 7: 46
57. Goldfinger G (1949) J Polym Sci 4: 93
58. Edwards A, Goldfinger G (1955) J Polym Sci 16: 589
59. Speight JG, Kovacic P, Koch FW (1971) J Macromol Sci, Rev Macromol Chem 5: 275
60. Kocacic P, Kyriakis A (1963) J Am Chem Soc 85: 454
61. Kovacic P, Wu C (1960) J Polym Sci 47: 448
62. Kovacic P, Hsu LC (1966) J Polym Sci 4: 5
63. Yamamoto T, Yamamoto A (1977) Chem Lett 356
64. Yamamoto T, Hayashi Y, Yamamoto Y (1978) Bull Chem Soc Jpn 51: 2091
65. Elsenbaumer RL, Shacklette LW (1986) In: Handbook of Conducting Polymers. Marcel Dekker Inc, New York
66. Rehahn M, Schlüter AD, Wegner G, Feast WJ (1989) Polymer 30: 1054
67. Rehahn M, Schlüter AD, Wegner G, Feast WJ (1989) Polymer 30: 1060
68. Rehahn M, Schlüter AD, Wegner G (1990) Makromol Chem 191: 1991
69. Rulkens R, Schulze M, Wegner G (1994) Macromol Rapid Commun 15: 669
70. Vahlenkamp T, Wegner G (1994) Macromol Chem Phys 195: 1933
71. Remmers M, Schulze M, Wegner G (1996) Macromol Rapid Commun 17: 239
72. Vanhee S, Rulkens R, Lehmann U, Rosenauer C, Schulze M, Köhler W, Wegner G (1996) Macromolecules 29: 5136
73. Lauter U, Meyer WH, Wegner G (1997) Macromolecules 30: 2092
74. Percec V, Okita S, Weiss R (1992) Macromolecules 25: 1816
75. Percec V, Okita S, Bae J (1992) Polym Bull 29: 271
76. Percec V, Pugh C, Cramer E, Okita S, Weiss R (1992) Makromol Chem Macromol Symp: 54/55: 113
77. Qian X, Pena M (1995) Macromolecules 28: 4415
78. Percec V, Bae JY, Zhao M, Hill DH (1995) J Org Chem 60: 176
79. Percec V, Bae JY, Zhao M, Hill DH (1995) J Org Chem 60: 1066
80. Percec V, Zhao M, Bae JY, Hill DH (1996) Macromolecules 29: 3727

81. Grob MC, Feiring AE, Auman BC, Percec V, Zhao M, Hill DH (1996) Macromolecules 29: 7284
82. Colon I, Kelsey DR (1986) J Org Chem 51: 2627
83. Ueda M, Ichikawa F (1990) Macromolecules 23: 926
84. Ueda M, Seino Y, Sugiyama J (1993) Polym J 25: 1319
85. Phillips RW, Sheares VV, Samulski ET, DeSimone JM (1994) Macromolecules 27: 2354
86. Wang Y, Quirk RP (1995) Macromolecules 28: 3495
87. Reddinger JL, Reynolds JR (1997) Macromolecles 30: 479
88. Tour JM, Stephens EB (1991) J Am Chem Soc 113: 2309
89. Stephens EB, Kinsey KE, Davis JP, Tour J M (1993) Macromolecules 26: 3519
90. Kim YH, Webster OW (1990) J Am Chem Soc 112: 4592
91. Kim YH, Webster OW (1992) Macromolecules 25: 5561
92. Cassidy PE, Marvel CS (1972) Macromol Synth 4: 7
93. Frey DA, Hasegawa M, Marvel CS (1963) J Polym Sci, Part A 1: 2057
94. Gin DL, Avlyanov JK, MacDiarmid AG (1994) Synth Met 66: 169
95. Ballard DGH, Courtis A, Shirley IM, Taylor SC (1983) J Chem Soc, Chem Commun: 954
96. Ballard DGH, Courtis A, Shirley IM, Taylor SC (1988) Macromolecules 21: 294
97. McKean DR; Stille JK (1987) Macromolecules 20: 1787
98. Gin DL; Conticello VP, Grubbs RH (1992) J Am Chem Soc 114: 3167
99. Lockhardt TP, Comita PB, Bergman RG (1981) J Am Chem Soc 103: 4082
100. Lockhardt TP, Bergman RG (1981) J Am Chem Soc 103: 4091
101. John JA, Tour JM (1994) J Am Chem Soc 116: 5011
102. Wallow TI, Novak BM (1991) J Am Chem Soc 113: 7411
103. Child AD, Reynolds JR (1994) Macromolecules 27: 1975
104. Scherf U, Müllen K (1991) Makromol Chem, Rapid Commun 12: 489
105. Scherf U, Müllen K (1992) Polymer 33: 2443
106. Scherf U, Müllen K (1992) Macromolecules 25: 3546
107. Scherf U, Bohnen A, Müllen K (1992) Makromol Chem 193: 1127
108. Scherf U (1993) Synth Met 55: 767
109. Chmil K, Scherf U (1993) Makromol Chem, Rapid Commun 14: 217
110. Tour JM, Lamba JJS (1993) J Am Chem Soc 115: 4935
111. Lamba JJS, Tour JM (1994) J Am Chem Soc 116: 11723
112. Hörhold HH, Helbig M, Raabe D, Opfermann J, Scherf U, Stockmann R, Weib DZ (1987) Chem 27: 126
113. Bradley DDC (1987) J Phys D, Appl Phys 20: 1389
114. Prasad PN (1991) Polymer 32: 1746
115. Bradley DDC (1992) Adv Mater 4: 756
116. Murase I, Ohnishi T, Noguchi T, Hirooka M (1984) Polym Commun 25: 327
117. Antoun S, Gagnon DR, Karasz FE, Lenz RW (1986) J Polym Sci, Polym Lett 24: 503
118. Kaino T, Kobayashi H, Kubodera T, Kurihara T, Saito S, Tsutsui T, Tokito S (1989) Appl Phys Lett 54: 1619
119. Wung CJ, Pang Y, Prasad PN, Karasz FE (1991) Polymer 32: 605
120. Shim HK, Hwang DH, Lee KS (1993) Makromol Chem 194: 1115
121. Shim HK, Lee KS, Jin JI (1996) Macromol Chem Phys 197: 3501
122. Burroughes JH, Bradley DDC, Brown AR, Marks RN, Mackay K, Friend RH, Burn PL, Holmes AB (1990) Nature 347: 539
123. Braun D, Heeger AJ (1991) Appl Phys Lett 58: 1982
124. Burn PL, Holmes AB, Kraft A, Bradley DDC, Brown AR, Friend RH, Gymer RW (1992) Nature 356: 47
125. Gustafsson G, Cao Y, Treacy GM, Klavetter F, Colaneri N, Heeger AJ (1992) Nature 357: 477
126. Greenham NC, Moratti SC, Bradley DDC, Friend RH, Holmes AB (1993) Nature 365: 628

127. Holmes AB, Bradley DDC, Brown AR, Burn PL, Burroughes JH, Friend RH, Greenham NC, Gymer RW, Halliday DA, Jackson RW, Kraft A, Martens JHF, Pichler K, Samuel IDW (1993) Synth Met 55-57: 4031
128. Yu G (1996) Synth Met 80: 143
129. Cornil J, Beljonne D, dos Santos DA, Shuai Z, Brédas JL (1996) Synth Met 78: 209
130. Yang Y, Pei Q, Heeger AJ (1996) J Appl Phys 79: 934
131. Lee CH, Yu G, Moses D, Sariciftci NS, Heeger AJ, Wudl F (1993) Phys Rev B 48: 15425
132. Martelock H, Griener A, Heitz W (1991) Makromol Chem 192: 967
133. Yu L, Bao Z (1994) Adv. Mater 4: 156
134. McDonald RN, Campbell TW (1960) J Am Chem Soc 82: 4669
135. Hörhold HH, Opfermann J (1969) Makromol Chem 129: 105
136. Manecke VG, Zerpner D (1969) Makromol Chem 129: 183
137. Kossmehl VG, Härtel M, Manecke G (1970) Makromol Chem 131: 37
138. Gourley KD, Lillya CP, Reynolds JR, Chien JCW (1984) Macromolecules 17: 1025
139. Huang WS, Jen KY, Angelopoulos AG, MacDiarmid AG, Cava MP (1990) Mol Cryst Liq Cryst 189: 237
140. Lenz RW, Handlovits CE (1960) J Org Chem 25: 813
141. Funke VW, Schütze EC (1963) Makromol Chem 74: 71
142. Debord D, Golé J (1971) Bull Chem Soc 4: 1401
143. Hörhold HHZ (1972) Chem 12: 41
144. Lahti PM, Sarker A, Garay RO, Lenz RW, Karasz FE (1994) Polymer 35: 1312
145. Moratti SC (1995) Synth Met 71: 2117
146. Baigent DR, Hamer PJ, Friend RH, Moratti SC, Holmes AB (1995) Synth Met 71: 2175
147. Hoeg DF, Lusk DI, Goldberg EP (1964) J Polym Sci Part B, Polym Lett 2: 697
148. Gilch HG, Wheelwright WL (1966) J Polym Sci Part A, Polym Chem 4: 1337
149. Hörhold HH, Opfermann J (1970) Makromol Chem 131: 105
150. Rajaraman L, Balasubramanian M, Nanjan MJ (1980) Curr Sci 49: 101
151. Cataldo F (1991) Polym Commun 32: 354
152. Rehahn M, Schlüter AD (1990) Macromol Chem, Rapid Commun 11: 375
153. Kumar A, Eichenger BE (1992) Macromol Chem, Rapid Commun 13: 311
154. Thorn-Csányi E, Pflug KP (1993) Macromol Chem, Rapid Commun 14: 619
155. Thorn-Csányi E, Kraxner P (1995) Macromol Chem, Rapid Commun 16: 147
156. Wessling RA, Zimmerman RG (1985) J Poly Sci, Polym Symp 72: 55
157. Lenz RW, Han CC, Stenger-Smith JD, Karasz FE (1988) J PolymSCi, Polym Chem Ed 26: 3241
158. Stenger-Smith JD, Lenz RW, Wegner G (1989) Polymer 30: 1048
159. Burn PL, Bradley DDC, Brown AR, Friend RH, Holmes AB (1991) Synth Met 41-43: 261
160. Schlenoff JB, Wang LJ (1991) Macromolecules 24: 6653
161. Berdeen A, Vanderzande D, Gelan J (1992) Synth Met 52: 387
162. Garay RO, Baier U, Bubeck C, Müllen K (1993) Adv Mater 5: 561
163. Murase I, Ohnishi T, Noguchi T, Hirooka M (1985) Polym Commun 26: 362
164. Antoun S, Karasz FE, Lenz RW (1988) J Polym Sci A, Polym Sci 26: 1809
165. Lenz RW, Han CC, Lux M (1989) Polymer 30: 1041
166. Jen KY, Shacklette LW, Elsenbaumer RL (1987) Synth Met 22: 179
167. Jin JI, Park CK, Shim HK (1993) Macromolecules 26: 1799
168. Shi S, Wudl F (1990) Macromolecules 23: 2119
169. Louwet F, Vanderzande D, Gelan J (1995) Synth Met 69: 509
170. Louwet F, Vanderzande D, Gelan J, Mullens J (1995) Macromolecules 28: 1330
171. Conticello VP, Gin DL, Grubbs RH (1992) J Am Chem Soc 114: 9708
172. Pu L, Wagaman MW, Grubbs RH (1996) Macromolecules 29: 1138
173. Wagaman MW, Grubbs RH (1997) Macromolecules 30: 3978
174. Heck RF (1968) J Am Chem Soc 90: 5518
175. Heck RF (1979) Acc Chem Res 12: 146
176. Heck RF. (1982) Org React 27: 345

177. de Meijere A, Meyer FE (1994) Angew Chem, Int Ed Engl 33: 2379
178. Greiner A, Heitz A (1988) Makromol Chem, Rapid Commun 9: 581
179. Heitz W (1988) Makromol Chem 189: 119
180. Brenda M, Greiner A, Heitz W (1990) Makromol Chem 191: 1083
181. Nishide H, Kaneko T, Nii T, Katoh K, Tsuchida E, Yamaguchi K (1995) J Am Chem Soc 117: 548.
182. Nishide H, Kaneko T, Nii T, Katoh K, Tsuchida E, Lahti P (1996) J Am Chem Soc 118: 9695
183. Nishide H, Hozumi Y, Nii T, Tsuchida E (1997) Macromolecules 30: 3986
184. Bäuerle P (1993) Adv Mater 5: 879
185. Kovacic P, McFarland KN (1979) J Poly Sci, Polym Chem Ed 17: 1963
186. Yamamoto T, Sanechika K, Yamamoto A (1980) J Polym Sci, Polym Lett Ed 18: 9
187. Lin J, Dudek LP (1980) J Polym Sci, Polym Chem Ed 18: 2869
188. Kossmehl G, Chatzitheodorou G (1981) Makromol Chem Rapid Commun 2: 551
189. Diaz AF (1981) Chem Scripta 17: 142
190. Tourillon G, Garnier F (1982) J Electroanal Chem 135: 173
191. Sugimoto R, Takeda S, Gu HB, Yoshino K (1986) Chem Express 1: 635
192. Yoshino K, Nakajima S, Fuji M, Sugimoto R (1987) Polym Commun 28: 309
193. Waltman RJ, Bargon J, Diaz AF (1983) Phys Chem 87: 1458
194. Roncali J, Lemaire M, Garreau R, Garnier F (1987) Synth Met 18: 139
195. Corradi R, Armes SP (1997) Synth Met 84: 453
196. Bao Z, Chan WK, Yu L (1995) J Am Chem Soc 117: 12426
197. Fanta PE (1964) Chem Rev 64: 613
198. Fanta PE (1974) Synthesis 9
199. Pomerantz M, Yang H, Cheng Y (1995) Macromolecules 28: 5706
200. Yamamoto T, Morita A, Miyazaki Y, Maruyama T, Wakayama H, Zhou Z, Nakamura Y, Kanbara T, Sasaki S, Kubota K (1992) Macromolecules 25: 1214
201. Yamamoto T, Wakabayashi S, Osakada K (1992) J Organomet Chem 428: 223
202. Miyazaki Y, Kanbara T, Osakada K, Yamamoto T (1993) Chem. Lett: 415
203. Masuda H; Kaeriyama K (1992) Makromol Chem, Rapid Commun 13: 461
204. Ueda M, Miyaji Y, Ito T, Oba Y, Sone T (1991) Macromolecules 24: 2694
205. Lemaire M, Delabouglise D, Garreau R, Guy A, Roncali J (1988) J Chem Soc, Chem Commun: 658
206. Sato M, Tanaka S, Kaeriyama K (1986) J Chem Soc, Chem Commun: 873
207. Jen KY, Miller GG, Elsenbaumer RL (1986) J Chem Soc, Chem Commun: 1346
208. Hotta S, Rughooputh SDV, Heeger AJ, Wudl F (1987) Macromolecules 20: 212
209. Tamao K, Sumitani K, Kumada M (1972) J Am Chem Soc 94: 4374
210. Gronowitz S (1985) The Chemistry of Heterocyclic Compounds: Thiophene and Its Derivatives. Wiley, New York,
211. Roncali J, Garreau R, Yassar A, Marque P, Garnier F (1987) J Phys Chem 91: 6706
212. Roncali J, Marque P, Garreau R, Garnier F, Lemaire M (1990) Macromolecules 23: 1347
213. Roncali J, Garreau R, Delabouglise D, Garnier F, Lemaire M (1989) J Chem Soc, Chem Commun: 679
214. Heywang G, Jonas F (1992) Adv Mater 4: 116
215. Li HS, Garnier F, Roncali J (1992) Macromolecules 25: 6425
216. McCullough RD, Williams SP (1993) J Am Chem Soc 115: 11608
217. Kang TJ; Kim JY, Kim KY, Lee C, Rhee SB (1995) Synth Met 69: 377
218. Lee C, Kim KY, Rhee SB (1995) Synth Met 69: 295
219. Middlecoff JS, Collard DM (1997) Synth Met 84: 221
220. Büchner W, Garreau R, Lemaire M, Roncali J (1990) J Electroanal Chem 277: 355
221. El Kassmi A, Büchner W, Fache F, Lemaire M (1992) J Electroanal Chem 326: 357
222. Ritter SK, Noftle RE, Ward AE (1993) Chem Mater 5: 752
223. Robitaille L, Leclerc M (1994) Chem Mater. 27: 1847

224. Kotkar D, Joshi V, Gosh K (1988) J Chem Soc, Chem Commun: 917
225. Sato M, Tanaka S, Kaeriyama K (1989) Makromol Chem: 1233
226. Lemaire M, Garreau R, Delabouglise D, Korri Youssoufi H, Garnier F (1990) New J Chem 14: 359
227. Patil AO, Ikenoue Y, Wudl F, Heeger AJ (1987) J Am Chem Soc 109: 1858
228. Ikenoue Y, Saida Y, Kira M, Tomozowa H, Yashima H, Kobayashi M (1990) J Chem Soc, Chem Commun: 1694
229. Chen SA, Hua MY (1993) Macromolecules 26: 7108
230. Arroyo-Villan MI, Diaz-Quijada GA, Abdou MSA, Holdcroft S (1995) Macromolecules 28: 975
231. Leclerc M, Diaz FM, Wegner G (1989) Makromol Chem 190: 3105
232. Sato M, Morii H (1991) Polym Commun 32: 42
233. Mao H, Holdcroft S (1992) Macromolecules 25: 554
234. Souto Maior RM, Hinkelmann K, Eckert H, Wudl F (1990) Macromolecules 23: 1268
235. Mao, H.; Xu, B.; Holdcroft, S. Macromolecules 1993, 26, 1163
236. Anderson MR, Selse D, Berggren M, Jarvinen H, Hjertberg T, Inganäs O, Wennerström O, Österholm JE (1994) Macromolecules 27: 6503
237. Sato M, Morii H (1991) Macromolecules 24: 1196
238. McCullough RD, Lowe RD (1992) J Chem Soc Chem Commun: 70
239. Gallazzi MC, Castellani L, Zerbi G, Sozzani P (1991) Synth Met 41-43: 495
240. McCullough RD, Lowe RD, Jayaraman M, Anderson DL (1993) J Org Chem 58: 904
241. Chen TA, Rieke RD (1992) J Am Chem Soc 114: 10087
242. Chen TA, O'Brien RA, Rieke RD (1993) Macromolecules 26: 3462
243. Chen TA, Rieke RD (1995) J Am Chem Soc 117: 233
244. Wu X, Chen TA, Rieke RD (1995,) Macromolecules 28: 2101
245. a) Jonas F, Heywang G, Schidtberg W (1989) Ger Offen DE 3, 813: 589,
246. Jonas F, Heywang G, Schidtberg W, Heinze J, Dietrich M (1989) Eur Pat App EP 339: 340
247. Jonas F, Heywang G, Schidtberg W, Heinze J, Dietrich M (1991) U.S. Patent No. 5,035: 926,
248. Heywang G, Jonas F (1992) Adv Mater 4: 116
249. Pei Q, Zuccarello G, Ahlskog M, Inganäs O (1994) Polymer 35: 1347
250. Gustafsson JC, Liedberg B, Inganäs O (1994) Solid State Ionics 69: 145
251. Dietrich M, Heize J, Heywang G, Jonas F (1994) J Electroanal Chem 369: 87
252. Kumar A, Reynolds JR (1996) Macromolecules29: 8551
253. Stéphan O, Schottland P, Le Gall PY, Chevrot C, Mariet C, Carrier M (1998) J Electroanal Chem443: 217
254. Sankaran B, Reynolds JR (1997) Macromolecules 30: 2582
255. Havinga EE, Mutsaers CMJ, Jenneskens LW (1996) Chem Mater 8: 769
256. Kumar A, Welsh DM, Morvant MC, Piroux F, Abboud KA, Reynolds JR (1998) Chem Mater 10: 896
257. Sotzing GA, Reynolds JR (1995) J Chem Soc, Chem Commun: 703
258. Sotzing GA, Reynolds JR, Steel PJ (1996) Chem Mater 8: 882
259. Reddinger JL, Sotzing GA, Reynolds JR (1996) J Chem Soc, Chem Commun: 1777
260. Sotzing GA, Reddinger JL, Katritzky AR, Soloducho J, Musgrave R, Steel PJ, Reynolds JR (1997) Chem Mater 9: 1578
261. Sapp SA, Sotzing GA, Reddinger JL, Reynolds JR (1996) Adv Mater 8: 808
262. Sotzing GA, Thomas CA, Reynolds JR, Steel PJ (1998) Macromolecules 31: 3750
263. Dall'Olio A, Dascola G, Varacca V, Bocchi V (1968) R Acad Sci C267: 433
264. Diaz AF, Kanazwa KK, Gardini GP (1979) J Chem Soc, Chem Commun: 635
265. Kanazawa KK, Diaz AF, Geiss RH, Gill WD, Kwak JF, Logan JA, Rabolt JF, Street GB (1979) J Chem Soc, Chem Commun: 854
266. Wynne KJ, Street GB (1985) Macromolecules 18: 2361
267. Diaz AF, Bargon J (1986) Handbook of Conducting Polymers. Skotheim TA (ed) Marcel Dekker Inc, New York, p. 81

268. Street, G. B. Handbook of Conducting Polymers T. A. Skotheim, Ed.; Marcel Dekker Inc.: New York, 1986, p. 265
269. Malhotra BD, Kumar N, Chandra S (1986) Progr Polym Sci 12: 179
270. Saunders BR, Fleming RJ, Murray KS (1995) Chem Mater 7: 1082
271. Novak P (1992) Electrochim Acta 37: 1227
272. Bartlett PN, Birkin PR (1993) Synth Met 61: 15
273. Salmon M, Kanazwa KK, Diaz AF, Krounbi M (1982) J Polym Sci, Polym Lett 20: 187
274. Chan HSO, Munro HS, Davies C, Kang ET (1988) Synth Met 22: 365
275. Mermilliod N, Tanguy J, Petiot F (1986) J Electrochem Soc 133: 6
276. Myers RE (1986) J Electron Mater 2: 61
277. Armes SP (1987) Synth Met 20: 365
278. Nishio K, Fujimoto M, Ando O, Ono H, Murayama T (1996) J Appl Electrochem 26: 425
279. Thiéblemont JC, Gabelle JL, Planche MF (1994) Synth Met 66: 243
280. Curran D, Gromshaw J, Perera SD (1991) Chem Soc Rev 20: 391
281. Deronzier A, Moutet JC (1996) Coord Chem Rev 147: 339
282. Havinga EE, van Horssen LW, ten Hoeve W, Wynberg H, Meijer EW (1987) Polym Bull 18: 277
283. Inagaki T, Hunter M, Yang XQ, Skotheim TA, Okaoto Y (1988) J Chem Soc, Chem Commun: 126
284. Rühe J, Ezquerra TA, Wegner G (1989) Synth Met 28: C177
285. Andrieux CP, Audebert P (1989) J Electroanal Chem 261: 443
286. Zotti G, Shiavon G, Berlin A, Pagani G (1989) Synth Met 28: C183
287. Martina S, Enkelmann V, Wegner G, Schlüter AD (1991) Synthesis: 613
288. Martina S, Enkelmann V, Wegner G, Schlüter AD (1992) Synth Met 51: 299
289. Groenendaal L, Peerlings HWI, van Dongen JLJ, Havinga EE, Vekemans JAJM, Meijer EW (1995) Macromolecules 28: 116
290. Brockmann TW, Tour JM (1994) J Am Chem Soc 116: 7435
291. Letheby H (1862) J Chem Soc 15: 161
292. Mohilner DM, Adams RN, Argersinger Jr WJ (1962) J Am Chem Soc 84: 3618
293. Jozefewicz M, Yu LT, Belorgey G, Buvet R (1967) J Polym Sci, Part C 16: 2931
294. Honzl J, Tlustakova M (1968) J Polym Sci, Part C 22: 451
295. Jozefewicz M, Yu LT, Perichon J, Buvet R (1969) J Polym Sci, Part C 22: 1187
296. MacDiarmid AG, Epstein AJ (1989) Faraday Discuss Chem Soc 88: 317
297. Epstein AJ (1991) Makromol Chem, Macromol Symp 51: 217
298. MacDiarmid AG (1997) Synth Met 84: 27
299. Green AG, Woodhead AE (1910) J Chem Soc Trans 97: 2388
300. Green AG, Woodhead AE (1912) J Chem Soc Trans 101: 1117
301. Chiang JC, MacDiarmid AG (1986) Synth Met 13: 193
302. Lu FL, Wudl F, Nowak M, Heeger AJ (1986) J Am Chem Soc 108: 8311
303. Lu FL, Allemand PM, Vachon DJ, Nowak M, Liu ZX, Heeger AJ (1987) J Am Chem Soc 109: 3677
304. Wang L, Jing X, Wang F (1986) Synth Met 13: 329
305. Scherr EM, MacDiarmid AG, Manohar SK, Masters JG. Sun Y, Tang X, Druy MA, Glatkowski PJ, Cajipe VB, Fisher JE, Cromack KR, Jozefowitcz ME, Ginder JM, McCall RP, Epstein AJ (1991) Synth Met 41-43: 735
306. Epstein AJ, Ginder JM, Zuo F, Bigelow RW, Tanner DB, Richter AF, Huang WS. MacDiarmid AG (1987) Synth Met 18: 303
307. Wang L, Jing X, Wang F (1991) Synth Met 41-43: 739
308. MacInnes D, Funt BL (1988) Synth Met 25: 235
309. Leclerc M, Guay J, Dao LH (1989) Macromolecules 22: 649
310. Pelter A, Jenkins I, Jones DE (1997) Tetrahedron 53: 10357
311. Gooding R, Lillya CP, Chien CW (1983) J Chem Soc, Chem Commun: 151
312. Jen KY, Maxfield M, Shacklette LW, Elsenbaumer RL (1987) J Chem Soc, Chem Commun: 309

313. Yamada S, Tokito S, Tsutsui T, Saito S (1987) J Chem Soc, Chem Commun: 1448.
314. Sarker A, Lahti PM, Karasz FE (1994) J Polym Sci, Part A 32: 65
315. Zhou Z, Maruyama T, Kanbara T, Iveda T, Ichimura K, Yamamoto T, Tokuda K (1991) J Chem Soc, Chem Commun: 1210
316. Reynolds JR, Ruiz JP, Child AD, Nayak K, Marynick DS (1991) Macromolecules 24: 678
317. Ruiz JP, Dharia JR, Reynolds JR (1992) Macromolecules 25: 849
318. Bao Z, Chan W, Yu L (1993) Chem Mater 5: 2
319. Niziurski-Mann R, Cava MP (1993) Adv Mater 5: 547
320. Parakka JP, Chacko AP, Nickles DE, Wang P, Hasegawa S, Maruyama Y, Metzger RM, Cava MP (1996) Macromolecules 29: 1928
321. Tour JM (1996) Chem Rev 96: 537
322. Bohnen A, Heitz W, Müllen K, Räder HJ, Schenk R (1991) Makromol Chem 192: 1679
323. Zotti G, Martina S, Wegner G, Schlüter AD (1992) Adv Mater 4: 798
324. Zerbi G, Veronelli M, Martina S, Schlüter AD, Wegner G (1994) Adv Mater 6: 385
325. Rohde N, Eh M, Geibler U, Hallensleben ML, Voigt B, Voigt M (1995) Adv Mater 7: 401
326. Nakayama J, Konishi T, Hoshino M (1988) Heterocycles 27: 1731
327. Tour JM, Wu R (1992) Macromolecules 25: 1901
328. Barbarella G, Zambianchi M, Di Toro R, Colonna Jr M, Iarossi D, Goldoni F, Bongini A (1996) J Org Chem 61: 8285
329. Emge A, Bäuerle P (1997) Synth Met 84: 213
330. Dodabalapur A, Katz HE, Torsi L, Haddon RC (1995) Science 269: 1560
331. Dodabalapur A, Torsi L, Katz HE (1995) Science 268: 270
332. Havinga EE, Rotte I, Meijer EW, ten Hoeve W, Wynberg H (1991) Synth Met 41-43: 473
333. Yasser A, Delabouglise D, Hmyene M, Nessak B, Horowitz G, Garnier F (1992) Adv Mater 4: 490
334. Garnier F (1994) Science 179: 1684
335. Reisch J, Schulte KE (1961) Angew Chem 7: 241
336. Schulte KE, Reisch J, Hörner L (1962) Chem Ber 7: 241
337. Stetter H, Schreckenberg S (1974) Chem Ber 107: 2453
338. Stetter H, Kuhlmann H (1991) Org React 40: 407
339. Scheeren J W, Oomes PHJ, Nivard RJF (1973) Synthesis: 149
340. Bracke W (1972) J Polym Sci, Part A-1 10: 975
341. Pouwer KL, Vries TR, Havinga EE, Meijer EW, Wynberg HJ (1988) Chem Soc, Chem Commun: 1432
342. ten Hoeve W; Wynberg H, Havinga EE, Meijer EW (1991) J Am Chem Soc 113: 5887
343. Tamao K, Yamaguchi S, Shiozaki M, Nakagawa Y, Ito Y (1992) J Am Chem Soc 114: 5867
344. Tamao K, Yamaguchi S, Shiro M (1994) J Am Chem Soc 116: 11715
345. Tamao K, Yamaguchi S, Ito Y, Matsuzaki Y, Yamabe T, Fukushima M, Mori S (1995) Macromolecules 28: 8668
346. Tamao K, Yamaguchi S (1996) Pure & Appl Chem 68: 139
347. Okinoshima H, Yamamoto K, Kumada M (1972) J Am Chem Soc 94: 9263
348. Deronzier A, Moutet JC (989) Acc Chem Res 1 22: 255
349. Wolf MO, Wrighton MS (1994) Chem Mater 6: 1526
350. Yamamoto T, Maruyama T, Zhou ZH, Ito T, Fukuda T, Yoneda Y, Begum F, Ikeda T, Sasaki S, Takezoe H, Fukuda A, Kubota K (1994) J Am Chem Soc 116: 4832
351. Peng Z, Yu L (1996) J Am Chem Soc 118: 3777
352. Sauvage JP (1990) Acc Chem Res 23: 321
353. Zhu S, Swager TM (1996) Adv Mater 8: 497
354. Zhu S, Swager TM (1996) J Am Chem Soc 118: 8713
355. Wang B, Wasielewski MR (1997) J Am Chem Soc 119: 12
356. Reddinger JL, Reynolds JR (1997) Macromolecules 30: 673

357. Reddinger JL, Reynolds JR (1998) Chem Mater 10: 1236
358. Reddinger JL, Reynolds JR (1998) Chem Mater 10: 3

Editor: Prof. G.Wegner
Received: September 1998

Diallyldimethylammonium Chloride and its Polymers

C. Wandrey*, J. Hernández-Barajas, D. Hunkeler

Laboratory of Polymers and Biomaterials, Department of Chemistry, Swiss Federal Institute of Technology, CH-1015 Lausanne, Switzerland
*e-mail: wandrey@igc.dc.epfl.ch

The pyrrolidimium structure resulting from the cyclopolymerization of the water soluble monomer diallyldimethylammonium chloride is present in a variety of advanced polymeric materials. These materials range from water soluble polyelectrolytes to highly ordered solids. Applied research on diallyldimethylammonium chloride has been performed in order to optimize the monomer and polymer syntheses, characterize the polymers produced, improve the material properties as well as the applied technologies, and to develop new products. In addition, fundamental research has included the study of polyelectrolyte behavior of diallyldimethylammonium chloride polymers in solution. This article comprehensively reviews the work on diallyldimethylammonium chloride summarizing the current knowledge and recent progress in the field of kinetics and mechanism of homo- and copolymer syntheses, chemical structures, polyelectrolyte behavior in solution, molecular characterization, and interactions in solution and at interfaces. In particular, peculiarities of the syntheses and characterization resulting from polyelectrolyte influences are discussed in detail. A variety of current and emerging applications are presented.

Keywords: Diallyldimethylammonium chloride, Polyelectrolytes, Polymerization, Characterization, Application

List of Symbols and Abbreviations 125

1   Introduction. 126

2   The Diallyldimethylammonium Chloride Monomer. 127

2.1   Syntheses. 127
2.2   Solution Properties 130

3   Poly(diallyldimethylammonium chloride) Structures. 132

3.1   Homopolymers 132
3.2   Copolymers 134

4   Polymerization of Diallyldimethylammonium Chloride 135

4.1   Homopolymerization. 135
4.1.1 Homogeneous Polymerization in Aqueous Solution 135
4.1.2 Heterophase Polymerization. 141
4.1.3 Individual Constants and Comparison to Experimental Data. . . . 141
4.2   Copolymerization. 143

4.2.1	Copolymerization with Acrylamide	143
4.2.2	Copolymerization with other Monomers	148
5	**Solution Properties of Diallyldimethylammonium Chloride Polymers**	**150**
5.1	Fundamentals	150
5.1.1	Polyelectrolyte Models and Counterion Condensation	150
5.1.2	Concentration Regimes	151
5.1.3	Interactions Between Polyion and Counterions	152
5.1.4	Electrolytic Conductivity	152
5.2	Experimental Results	153
5.2.1	Counterion Activity	153
5.2.2	Electrolytic Conductivity	156
6	**Molecular Characterization of Diallyldimethylammonium Chloride Polymers**	**164**
6.1	General Considerations	164
6.2	Determination of Molar Masses	165
6.2.1	Homopolymers	165
6.2.2	Copolymers	166
6.3	Determination of Molar Mass Distributions	168
6.3.1	Chromatography	168
6.3.2	Other Methods	169
6.4	Determination of Structural Nonuniformities	169
7	**Interactions of Diallyldimethylammonium Chloride Polymers in Solution and at Interfaces**	**170**
7.1	Interactions with Low Molecular Weight Components	170
7.2	Interactions with Oppositely Charged Polymers	171
7.3	Interactions at Surfaces	172
8	**Applications of Diallyldimethylammonium Chloride Homo- and Copolymers**	**172**
8.1	Paper Manufacturing	173
8.1.1	Retention and Drainage Agents	173
8.1.2	Wet Strength Additives	173
8.1.3	Other Paper Manufacturing Agents	174
8.2	Mining Industry	174
8.3	Water Treatment Industry	175
8.4	Miscellaneous Applications	176
9	**Conclusions**	**176**
10	**References**	**177**

List of Symbols and Abbreviations

AAM	acrylamide
ADMA	allyldimethylamine
AOT	sodium di-2-ethyl-hexyl sulfosuccinate
APS	ammonium persulfate
DADMAC	diallyldimethylammonium chloride
DAMA	diallylmethylamine
DMA	dimethylamine
KMHS	Kuhn-Mark-Houwink-Sakurada relationship
LS	light scattering
MALLS	multi angle laser light scattering
MEAC	2-(methacryloyloxy)ethyltrimethylammonium chloride
MTAAC	methyltriallylammonium chloride
NaPSS	poly(sodium styrene sulfonate)
NMVA	N-methyl-N-vinylacetamide
OS	osmometry
PAAM	poly(acrylamide)
PDADMAC	poly(diallyldimethylammonium chloride)
PEL	polyelectrolyte
SEC	size exclusion chromatography
SMO	sorbitan monooleat
TEA	triethanolamine
UC	ultracentrifugation
A_X	virial coefficient of the Xth order
a	monomer unit length
a_c	counterion activity
b	charge distance
C_M	monomer transfer constant
c	concentration in mol L^{-1}
c_p	polyelectrolyte concentration in monomol L^{-1}
c_s	salt concentration in mol L^{-1}
c^*, c_b^*, c^{**}	critical concentrations in monomol L^{-1}
D_i/D_i°	self diffusion parameter
f	initiator effectivity
f_a	counterion activity coefficient
f_c	conductivity coefficient
f_o	osmotic coefficient
I	initiator
k	overall rate constant
k_d	rate constant of initiator decomposition
k_{deg}	rate constant of chain degradation
k_p	rate constant of propagation
k_t	rate constant of termination
L	contour length

L_p	persistence length
l_B	Bjerrum length
l_D	Debye length
M	monomer
M_n	number-average molar mass
M_w	weight-average molar mass
M_z	z-average molar mass
N_A	Avogadros' number
P_c	cyclized polymer radical
R_i	initiation rate
R_g	radius of gyration
R_p	overall polymerization rate
R_t	termination rate
r_X	reactivity ratio of the monomer X
η	viscosity
$[\eta]$	intrinsic viscosity
κ	specific conductance
Λ	equivalent conductivity
$\Lambda°$	equivalent conductivity at infinite dilution
Λ_{max}	maximum equivalent conductivity
λ_X	equivalent conductivity of the ion X
$\lambda°_X$	equivalent conductivity of the ion X at infinite dilution
ξ	charge density parameter (Manning parameter)

1
Introduction

The past five decades have been characterized by the rapid developments in the field of macromolecular chemistry. This has led, as is well known, to a broader application of synthetic polymeric materials. Due to the increasing requirements of environmental and health protection, processes and applications involving water treatment have become more regulated. Therefore, the development of technologies for water treatment and waste water processing has become necessary for the protection of the aqueous resources. One group of specialty polymers which has become significant, due to the development of auxiliary materials, additives and finishing components, are the synthetic anionic, cationic and ampholytic water soluble polymers or polyelectrolytes. Among these, polymeric quaternary ammonium compounds have historically been the most important and extensively used cationic polyelectrolytes. The first quaternary ammonium polymers of technical interest were synthesized from diallyldimethylammonium chloride (DADMAC) [1-4]. Although the polymer was first prepared in the 1950s [5], and the kinetics and mechanism of the polymerization process [6-16] as well as the structure of the resulting polyelectrolytes were elucidated in the eighties [17-19], interest in research on poly(diallyldimethylammonium chloride) (PDADMAC) has not diminished. On one hand, this is based on its unique chemical structure and on the other hand on its versatile applicability [1, 2, 15, 20–27]. PDADMAC possesses a backbone of cyclic units resulting from the special cyclopolymerization of DADMAC [1, 3]. Additionally, the highly hydrophilic permanently charged quaternary ammonium groups provide the polymer with a high water solubility and solution properties correspond to those of strong polyelectrolytes [27, 28]. It should be mentioned that it was the first polymer to be approved by the U.S. Food and Drug Administration for the use in potable water treatment [29]. Although the primary industrial applications remain in flocculation, dewatering, coagulation, retention, flotation and similar separation processes, other emerging applications have been reported in recent years. These will be detailed in Sect. 8. Furthermore, new materials have been produced via copolymerization with ionic or nonionic monomers. The syntheses of new structures and their broad use has strongly influenced the development and improvement of characterization methods as well as the basic investigation of the physical solution properties of these polyelectrolytes.

This article will summarize results and information derived from basic and applied research on DADMAC and its polymers. Contrarily to other specific publications, this review will include discussions of the synthesis, chemical structure, molecular characterization, polyelectrolyte behavior, complex formation, and applications. It will be shown that the real solution behavior of polyelectrolytes cannot be investigated separately from their chemical structure and that it is essential to study synthesis and characterization of polyelectrolytes along with their physico-chemical properties.

2
The Diallyldimethylammonium Chloride Monomer

2.1
Syntheses

Diallyldimethylammonium chloride is exclusively synthesized from dimethylamine (DMA) and allylchloride, although other methods, such as the synthesis starting with diallyl aminocyanide, have been elaborated [2]. From the DMA process monomers varying in quality are produced. Therefore, the selection of the monomer synthesis procedure primarily depends on the desired purity of the final product. Generally, three qualities of DADMAC can be produced:
– solid diallyldimethylammonium chloride
– purified aqueous monomer solutions without sodium chloride
– aqueous monomer solutions containing sodium chloride

Solid DADMAC is synthesized from dimethylamine and allylchloride by a two step process (Fig. 1). The first step is the alkylation of dimethylamine with allylchloride in an aqueous alkaline medium. This step is followed by a quaternization in an organic medium again with allylchloride [30–32]. To prepare the pure solid the separation and purification of the intermediate product allyld-

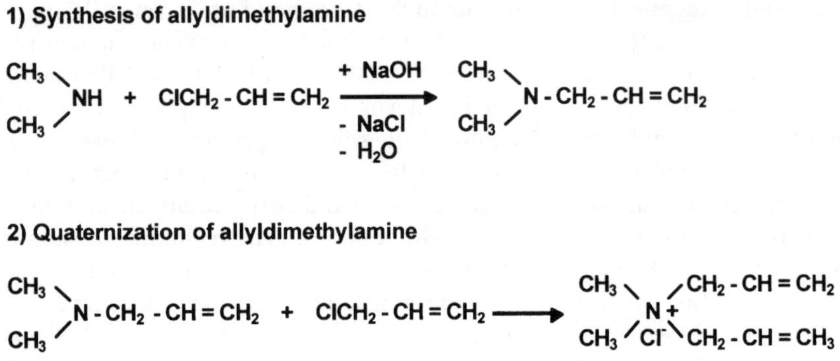

1) Synthesis of allyldimethylamine

$$\begin{array}{c} CH_3 \\ \diagdown \\ NH \\ \diagup \\ CH_3 \end{array} + ClCH_2\text{-}CH=CH_2 \xrightarrow[\substack{-NaCl \\ -H_2O}]{+NaOH} \begin{array}{c} CH_3 \\ \diagdown \\ N\text{-}CH_2\text{-}CH=CH_2 \\ \diagup \\ CH_3 \end{array}$$

2) Quaternization of allyldimethylamine

$$\begin{array}{c} CH_3 \\ \diagdown \\ N\text{-}CH_2\text{-}CH=CH_2 \\ \diagup \\ CH_3 \end{array} + ClCH_2\text{-}CH=CH_2 \longrightarrow \begin{array}{c} CH_3 CH_2\text{-}CH=CH_2 \\ \diagdown \diagup \\ N+ \\ \diagup Cl^- \diagdown \\ CH_3 CH_2\text{-}CH=CH_3 \end{array}$$

Fig. 1. Two step synthesis of diallyldimethylammonium chloride

imethylamine (ADMA) is necessary prior to quaternization [30–32]. Figure 2 shows the details of the synthesis of pure solid DADMAC.

The technical synthesis is normally carried out via a twofold alkylation of dimethylamine with allylchloride in aqueous alkaline medium [2, 33, 34]. Following purification this results in polymerizable monomer solutions. The extent of the purification determines the final monomer concentration and the residual content of sodium chloride. Normally, the monomer concentration of the resulting solutions is between 50 and 70% [2]. Depending on the purity of the raw materials and the reaction conditions the technical monomer solutions can contain small amounts of, for example, ADMA, diallymethylamine (DAMA), or methyltriallylammomium chloride (MTAAC) [35].

While DADMAC is primarily employed in homo- and copolymerization as a monomer to produce cationic or amphoteric polymers or gels, it is also a source for the synthesis of sulfobetains or sulfobetains with additional anionic groups by sulfocyclization, sulfocyclosulfonation, or sulfocyclosulfination [36]. Figure 3 summarizes these reactions.

The melting point of DADMAC was found to be 151–152.5 °C [32]. The crystalline product is very hygroscopic and soluble in water, alcohols, acetone, 1-methyl-2-pyrrolidone, tetramethyl urea, or dimethylformamide. ^{1}H-NMR analysis of the pure DADMAC shows the following signals: s 6.93 (N-CH$_3$); d 6.05 (N-CH$_2$-); m 3.53–4.50 (-CH=CH$_2$) ppm [32].

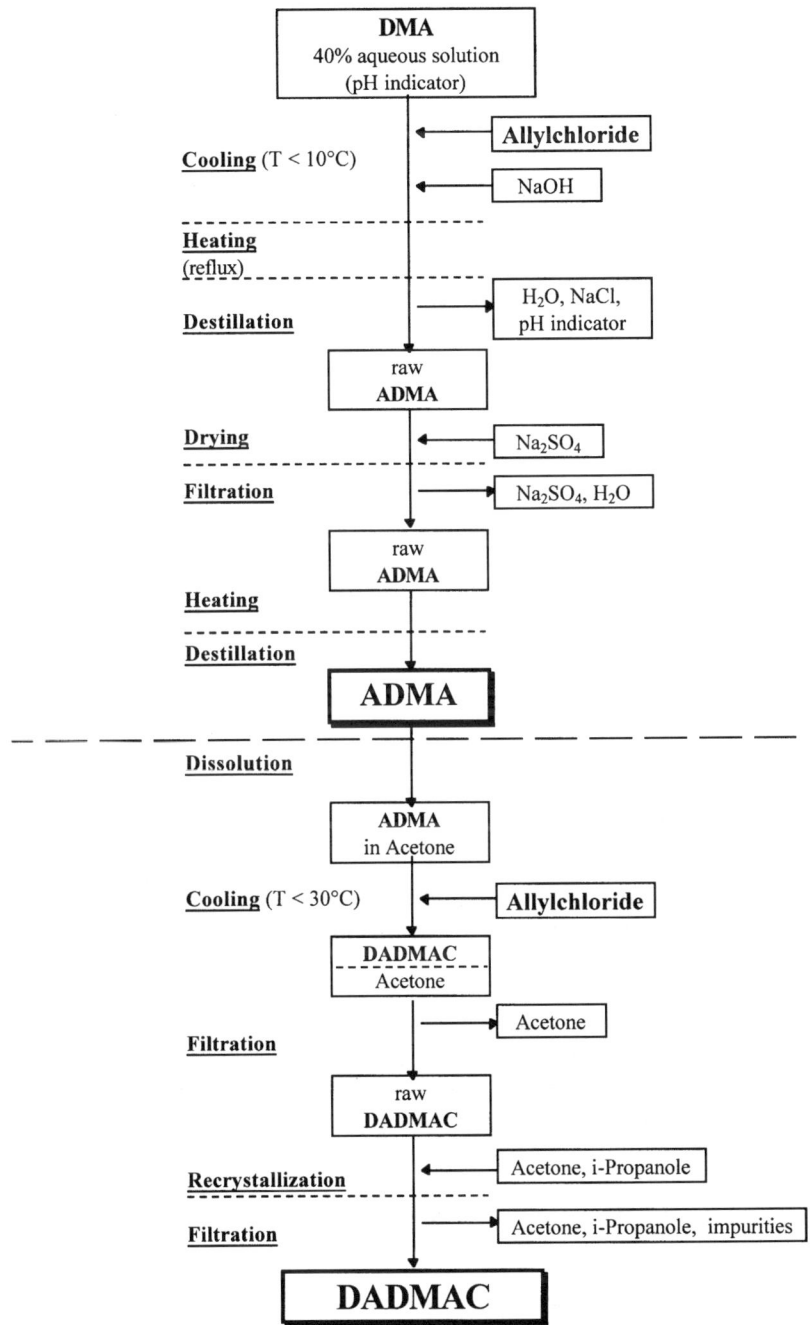

Fig. 2. Procedure for the synthesis of highly purified diallyldimethylammonium chloride. (DMA – dimethylamine, ADMA – allyldimethylamine)

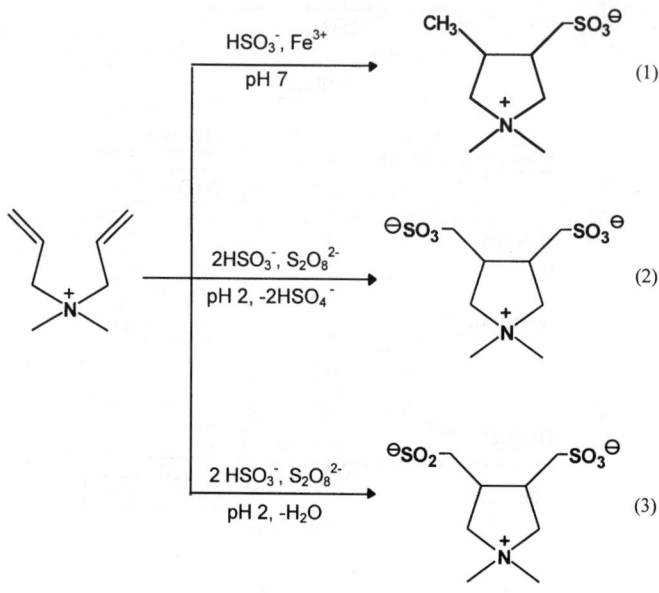

Fig. 3. Synthesis of sulfobetaines from diallyldimethylammonium chloride [36] (1: Sulfocyclization, 2: Sulfocyclosulfonation, 3: Sulfocyclosulfination)

2.2
Solution Properties

Density and Viscosity

The concentration dependence of both the density and the viscosity of the aqueous monomer solutions show an unusual curvature as indicated in Figs. 4 and 5.

In contrast to NaCl or tetramethylammonium bromide, also shown in Fig. 4, the concentration dependence of the density is less marked. However, the slopes of the density curves measured at 20 °C and 35 °C for DADMAC increase with the concentration. This indicates a change of the interaction with water is likely caused by the formation of ordered structures such as associates [32, 37]. The greatest change of the slope is located at approximately 1.5 mol L^{-1}. The influence of this monomer structure formation on the polymerization behavior will be discussed in Sect. 4. The non-linear concentration dependence of the viscosity is illustrated in Fig. 5. Here, a strong increase of this solution parameter is observed at approximately 1.5 mol L^{-1} indicating a change of intermolecular interactions [32, 37].

Electrochemical Solution Properties

The concentration dependencies of both the equivalent conductivity (Λ) and the chloride ion activity coefficient (f_a) of the monomer DADMAC are not different

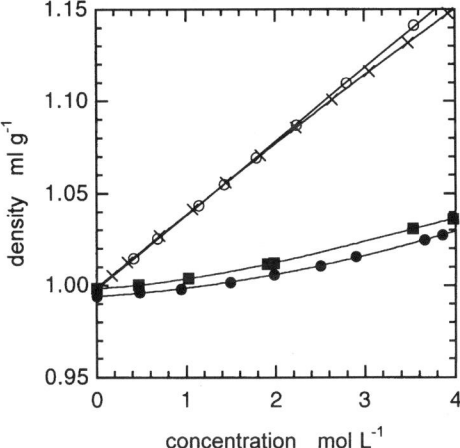

Fig. 4. Comparison of the concentration dependencies of the densities of aqueous salt solutions and DADMAC solutions (x NaCl, T=20 °C; ○ tetramethylammonium bromide, T=20 °C; ■ diallyldimethylammonium chloride, T=20 °C; ● diallyldimethylammonium chloride, T=35 °C) (Data taken from [32])

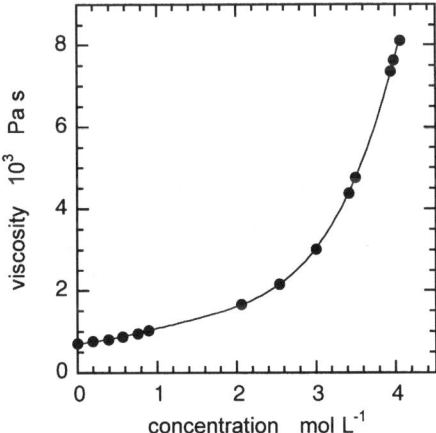

Fig. 5. Concentration dependence of the viscosity of the aqueous DADMAC solutions (T=35 °C) (Data taken from [32])

from those of the strong NaCl electrolyte. The chloride ion activity reaches a value of 1.0 at high dilutions [38]. This confirms that the quaternary ammonium group is a strong electrolyte. From conductivity measurements for DADMAC at 25 °C a limiting equivalent conductivity of $\Lambda°=111.1$ S·cm^2·mol^{-1} was calculated [38]. Assuming complete dissociation and taking the limiting equivalent conductivity of the chloride ion ($\lambda°_{chloride}=76.31$ S·cm^2·mol^{-1}) from [39], a value of

Table 1. Limiting equivalent conductivities ($\lambda°_{cation}$) of quaternary ammonium ions [39] (T=25 °C)

Ion	Structure	$\lambda°_{cation}$ [S·cm²·mol⁻¹]
ammonium	H_4N^+	73.6
tetramethylammonium	$C_4H_{12}N^+$	44.9
diallydimethylammonium	$C_8H_{16}N^+$	**34.8**
tetraethylammonium	$C_8H_{20}N^+$	32.7
tetrapropylammonium	$C_{12}H_{28}N^+$	23.4
tetrabutylammonium	$C_{16}H_{36}N^+$	19.5

$\lambda°_{cation}$=34.8 S·cm²·mol⁻¹ was obtained for the cation $C_8H_{16}N^+$. Data in Table 1 show that this limiting equivalent conductivity fits onto a single line with those of the other quaternary ammonium salts of the homologues series.

In Sect. 5 the concentration dependencies of Λ and f_a will be used for a comparison with curves determined from dilute polymer solutions. Therefore, they are given together with the polymer data in Figs. 14 and 18 of Sect. 5.

3
Poly (diallyldimethylammonium chloride) Structures

3.1
Homopolymers

Following the findings of Butler et al. [5], that diallyl quaternary ammonium halides form water soluble polymers, the structures of polymers produced via a ring-closing mechanism have been the subject of intensive research. A cyclic structure was elucidated and the mechanism was defined as an alternating intra-intermolecular chain propagation, later termed "cyclopolymerization" [3]. Based on the general scheme presented in Fig. 6 the chemical structure of PDADMAC is determined by
- the ring size of the cyclic units
- portions of cyclic and linear structure units
- extent of branching or crosslinking resulting from the reaction of the pendent double bond.

Ring Size

Originally, the formation of a six-membered ring was proposed [40], however, with the aid of ¹³C-NMR techniques it has been shown that PDADMAC consists exclusively of five-membered rings [17-19, 41]. Specifically, only configurational isomers of quaternary pyrrolidinium rings were detected and these provide a 6:1 ratio of *cis-* and *trans* configurations. The rings are connected by ethylene bridges in the 3,4-position [17, 19]. Interestingly variations in the conditions of the synthesis, for example, changes in solvent (water, acetone, 1-methyl-2-pyrrolid-

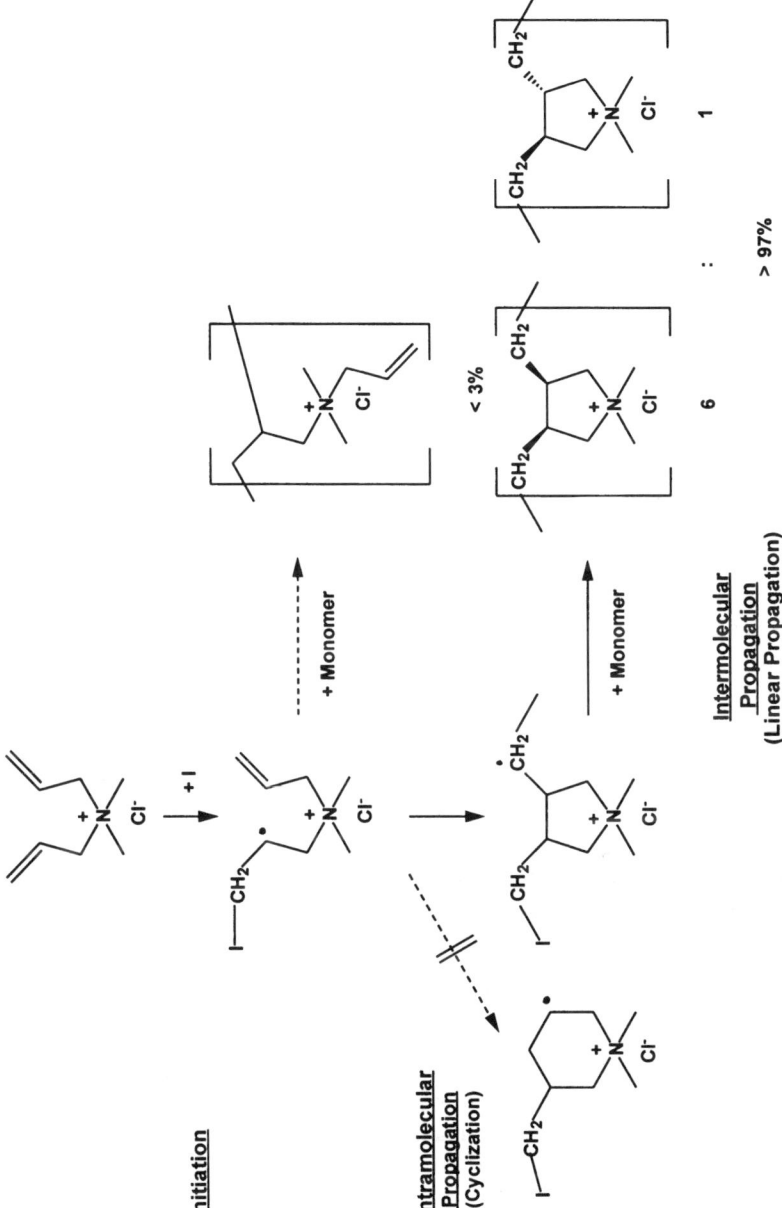

Fig. 6. Cyclopolymerization of diallyldimethyl ammonium chloride and chain structures of the resulting polymers

ine, tetramethylurea, dimethylformamide [19]), the initiation mechanism (radical with various initiators [17], anionic, radiation [19]), or the reaction temperature [17] do not influence the ring size and the 6:1 ratio of *cis-* and *trans* substitution [17, 19]. [1)]

Chain Architecture

PDADMAC obtained by the free radical polymerization with ammonium persulfate (APS) in water at low conversion contains only 0.1–3% pendent double bonds (1.5 mol L^{-1} <[M] <4.26 mol L^{-1}, 35 °C <T <60 °C), as detected by radiochemical determination of residual double bonds with ^{203}Hg-acetate following thin-layer chromatographic separation of the hydroxymercurated products [32, 42, 43]. The portion of noncyclized products increases with temperature and monomer concentration and decreases with conversion resulting in chain branching or crosslinking [19]. The extent of chain branching is influenced by the conditions of synthesis. It can be neglected (<1%) if the pure monomer is polymerized under mild conditions (low temperature, [M] <4 mol L^{-1}). However, under production conditions polymers with a higher degree of branching result, primarily due to the impurities of the technical monomer solution. Figure 7 shows possible repeat-unit structures if methyltriallylammomium chloride is present. It should be mentioned that as little as 3 mol % of MTAAC in the monomer solution produces completely crosslinked polymers which are insoluble in water [44].

3.2
Copolymers

^{13}C-NMR spectroscopy has also been used to investigate the composition of DADMAC copolymers [38, 45–47]. Copolymers with acrylamide (AAM) have been extensively studied. Based on the chemical shifts in the ^{13}C-NMR spectra of the homopolymers of DADMAC [17–19] and AAM [48–50], the copolymer spectra can be analyzed. Specifically from two likely diad structures (m/r) in PAAM and six different diad structures (r/m, c/t) in PDADMAC, eight different diad structures can be expected for the copolymers. A detailed NMR analysis has therefore been carried out in order to determine the copolymer compositions. The reactivity ratios obtained were found to be in good agreement with the results from other methods, such as potentiometric titration or elementary analysis, provided that the DADMAC in the copolymer was below approximately 70% [38].

---

1 Although the 5-membered ring of PDADMAC has been conclusively documented for over 15 years, the 6-membered ring structure is still cited by several authors [126, 211, 219, 220].

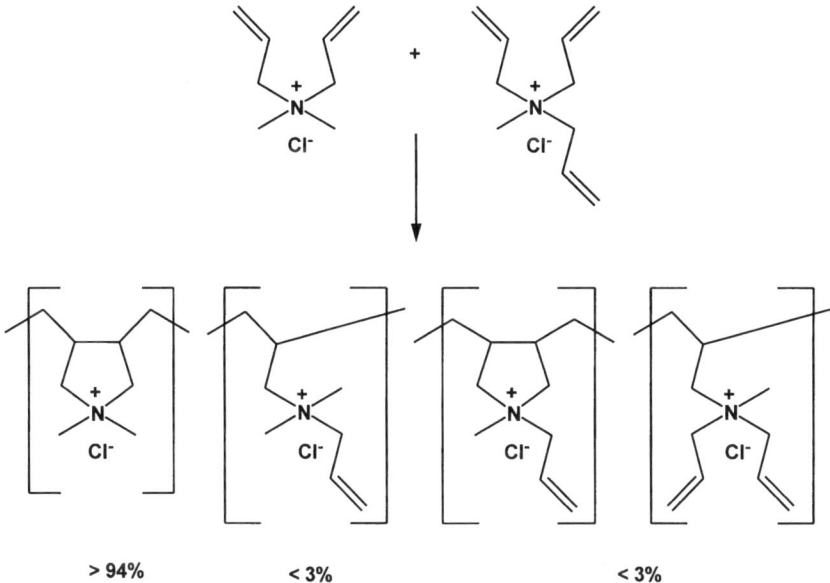

Fig. 7. Structural units of technical PDADMAC

4
Polymerization of Diallyldimethylammonium Chloride

4.1
Homopolymerization

The homopolymerization of DADMAC is possible in several organic solvents such as acetone, 1-methyl-2-pyrrolidone, tetramethylurea, or dimethylformamide. Various initiation methods including radical, ionic, or x-ray induced polymerization have been employed [19]. Since the monomer solubility is limited in these solvents, and the resulting homopolymer is soluble only in water, methanol and acidic acid, the polymerization in aqueous solutions are preferred. Polymerization in both homogeneous and heterogeneous systems have been studied and the kinetics and mechanisms were investigated in aqueous solution and in inverse-emulsion [6–16, 52, 53].

4.1.1
Homogeneous Polymerization in Aqueous Solutions

Overall Rate Equations

With various water soluble initiators, different overall polymerization rate equations were obtained. These equations are summarized in Table 2.

Table 2. Overall polymerization rate equations $R_p = k [I]^a [M]^b$ for the initiation with azo initiators (Azo), ammonium persulfate (APS), ammonium persulfate/triethanolamine (APS/TEA), ammonium persulfate/allyldimethylamine (APS/ADMA)

Initiator	a	b	k	Eq.	Ref.
Azo*	0.5	2	$4.25 \cdot 10^{-5}$ $L^{1.5} \cdot mol^{-1.5} \cdot s^{-1}$ (50 °C)	(1)	[10, 53]
APS	0.8	2.9	$7.25 \cdot 10^{-5}$ $[L^{2.7} \cdot mol^{-2.7} \cdot s^{-1}]$ (35 °C), $[M] > 1.5$ mol l^{-1}	(2)	[6, 32]
	0.47	2.3			[59]
APS/TEA	0.65 / 0.5 (pH >7)	2	f(pH)	(3)	[11]
APS/ADMA	0.5 / 0.25 (pH >7)	2	f(pH)	(4)	[11]

* 4,4'-azobis(4-cyanopentanoic acid)

Using APS initiator, for low monomer concentrations ([M] <1 mol L^{-1}), deviations from the ideal overall kinetics were not observed [52, 54]. At higher monomer concentrations, particularly at [M] >1.5 mol L^{-1}, marked deviations from the undisturbed radical polymerization could be identified. Figure 8 illustrates this unusual polymerization behavior for the APS initiated solution polymerization. The monomer exponent increases with the DADMAC concentration, reaching a constant value for monomer concentrations greater than 1.5 mol L^{-1} [6].

By calculating the overall rate constant k in Eq. (2) from experimental R_p data over monomer concentrations, the change in polymerization behavior could be identified to occur below 1.5 mol L^{-1} of the monomer [32]. This is demonstrated in Fig. 9.

Kinetic and Mechanistic Peculiarities

Deviations from the ideal radical elementary reaction scheme have been examined and various kinetic models were proposed [6, 7, 11, 16, 51]. Extended kinetic investigations [6–12, 32, 53, 54] have revealed significant kinetic and mechanistic anomalies:

- the formation of ionic complexes of the cationic monomer with anionic peroxide initiators [10] whereby the monomer-initiator complexes decompose faster than unbond initiators [10, 11, 55]
- a linear dependence of the ratio $k_p/k_t^{0.5}$ on the monomer concentration (Fig. 10) [6]

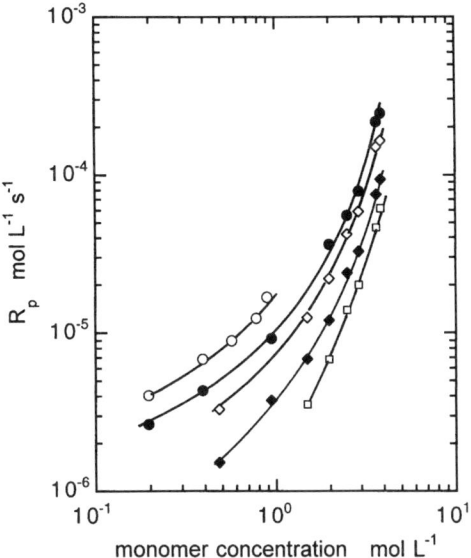

Fig. 8. Solution polymerization of DADMAC. Influence of the monomer concentration on the overall polymerization rate ([I]: ○ $5 \cdot 10^{-2}$; ● $3 \cdot 10^{-2}$; ◇ $2 \cdot 10^{-2}$; ◆ $1 \cdot 10^{-2}$; □ $5 \cdot 10^{-3}$ mol L^{-1}; T=35 °C) (Data taken from [6])

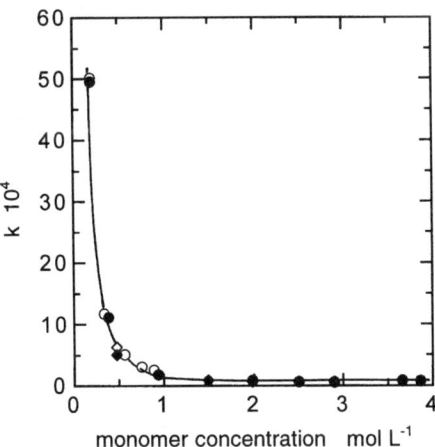

Fig. 9. Solution polymerization of DADMAC. Influence of the monomer concentration on the overall rate constant k in Eq. (2) ([I]: ○ $5 \cdot 10^{-2}$; ● $3 \cdot 10^{-2}$; ◇ $2 \cdot 10^{-2}$; ◆ $1 \cdot 10^{-2}$; □ $5 \cdot 10^{-3}$ mol L^{-1}; T=35 °C) (Data taken from [6])

- chain termination by combination of cyclized polymer radicals with chlorine atoms which result from the redox reaction of persulfate ions with monomer chloride ions [15]

The increase of $k_p / k_t^{0.5}$ has been the topic of debate [6, 8, 9, 11, 16, 56–58]. Hahn, Jaeger and Wandrey have assumed a structure/associate formation of monomer cations as well as the reduced electrostatic repulsion between the growing polymer cation radical and the monomer cation with increasing ionic strength to be the reason for this increase [6, 11, 16, 56]. However, Topchiev et al. discussed this result in terms of the dependence of the termination reaction on the initial monomer solution viscosity [8, 9, 57]. The former assumption is supported by the concentration dependence of the density of the monomer solution (Fig. 4) pointing to structure formation, and by the knowledge of the electrostatic screening if a low molecular weight salt is added to a polyelectrolyte solution [27, 28]. The latter will be discussed extensively in Sect. 5.

During the polymerization, DADMAC acts not only as a monomer, but also as a low molecular weight electrolyte which suppresses the Coulombic interactions. The addition of neutral low molecular salts (NaCl [10], NaBr [52], tetramethylammonium chloride [13]) leads in the same manner to a polymerization rate increase. Even though the viscosity of the monomer solution increases with the concentration (Fig. 5), it should be taken into account that the viscosity in the polymerizing system is more strongly influenced by the polymer than by the monomer, even at low conversions [51].

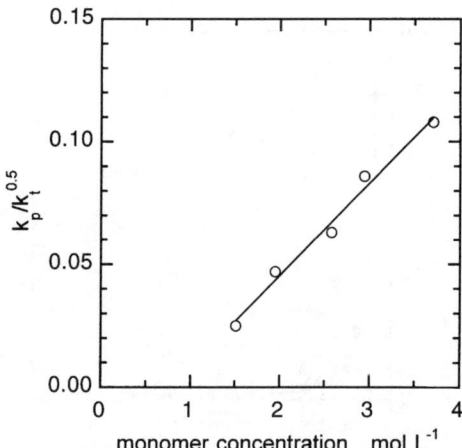

Fig. 10. Solution polymerization of DADMAC. Dependence of the rate constant ratio $k_p/k_t^{0.5}$ on the monomer concentration assuming termination by combination (T=35 °C) (Data taken from [6])

Kinetic Models

Table 3 summarizes all reactions of the APS initiated polymerization of DADMAC.

The individual side reactions (Table 3) result in the overall rate equation of persulfate initiation [10]:

$$R_i = 2fk_d \left[S_2O_8^{2-} \right]^{1.5} \left[M^+ \right] \left[Cl^- \right] \tag{5}$$

the rate equation for propagation [6,7]:

$$R_p = k_p \left[P_c \bullet \right] \left[M \right]^2 \tag{6}$$

and the rate equation for termination [10]:

$$R_t = k_t \left[P_c \bullet \right]^2 + k_{\text{deg}} \left[P_c \bullet \right] \left[Cl \bullet \right] \tag{7}$$

which includes both termination by recombination of two cyclized polymer radicals $P_c\bullet$ and the termination with chlorine atoms.

Under steady-state conditions, Equations 5–7 can be reduced to yield the overall rate equation of the APS initiated aqueous solution polymerization of DADMAC [10]:

$$R_p = \left(\frac{2fk_d k_p^2}{k_t} \right)^{0.5} [I]^{0.75} \left[M^+ \right]^{2.5} \left[Cl^- \right]^{0.5} - \frac{k_{\text{deg}} k_p}{2k_t} \left[Cl \bullet \right] \left[M \right]^2 \tag{8}$$

where the first term reflects the influence of the complex initiation and the specific propagation. Since normally $[M^+]=[Cl^-]$, the power with respect to [M] becomes third order. The second term represents the termination by the chlorine atoms. The side reactions of the monomer also find their expression in the monomer transfer constant determined for APS initiation, $C_M(APS)=2.5 \cdot 10^{-3}$ [6]. This value is much higher than the appropriate constant for azo initiation, $C_M(Azo)=1 \cdot 10^{-4}$ [10].

For the initiation by azo initiators only the dependence $k_p / k_t^{0.5}=f([M])$ has to be considered in a kinetic model [10]. Accordingly, an initiator exponent of 0.5 and a monomer exponent of 2 are valid. By adding amine the decomposition velocity of APS is increased by an orders of magnitude. The chain side reactions with the monomer and termination by chlorine atoms are then significantly suppressed which results in a monomer exponent of 2 and higher molar masses of the homopolymer [11]. The kinetics of 2.3 order in monomer and 0.47 order in initiation [59], explained by partial cyclization and termination of cyclized radicals, could not be confirmed.

Table 3. Reaction scheme of the APS initiated homopolymerization of DADMAC

Polymerization step	Chemical reaction	Reaction type
Primary radical formation	$S_2O_8^{2-} + 2\,M^+ \rightarrow [M^+\cdots S_2O_8^{2-}\cdots M^+] \rightarrow 2\,SO_4^{-}\bullet + 2\,M^+$	Complex formation
	$S_2O_8^{2-} + 2\,Cl^- \rightarrow SO_4^{-}\bullet + Cl\bullet + SO_4^{2-}$	Redox reaction
Initiation	$SO_4^{-}\bullet + M \rightarrow SO_4 M\bullet$	Chain start
	$Cl\bullet + M \rightarrow ClM\bullet$	Chain start
Propagation	$P_c\bullet + M \rightarrow P_l\bullet$	Linear propagation
	$P_l\bullet \rightarrow P_c\bullet$	Cyclization
Termination	$2\,P_c\bullet \rightarrow P_c$	Termination by combination
	$P_c\bullet + Cl\bullet \rightarrow P_c Cl$	Chain degradation
	$P_c\bullet + M \rightarrow P_c + M\bullet$	Chain transfer

4.1.2
Heterophase Polymerization

The initial rate of polymerization of the inverse emulsion polymerization using sodium di-2-ethyl-hexyl sulfosuccinate (AOT) and sorbitan monooleat (SMO) as emulsifiers and an oil soluble azo initiator can be expressed by [13]:

$$R_p = k[I]^{0.4}[AOT]^{0.5}[SMO]^{-0.4}[DADMAC]^3 \qquad (9)$$

with k=1.9 · 10^{-5} [L$^{2.5}$· s^{-1} · mol$^{-2.5}$] at 60.5 °C.

Based on experimental results the loci of polymerization are assumed to be the micelles and latex particles. The 3rd power with respect to monomer concentration in Eq. (9) results from the 2nd order polymerization reaction in aqueous solution as well as from the influence of the monomer concentration on the partition equilibrium of the monomer between micelles and monomer/water droplets [13]. This influence is shown in Fig. 11.

4.1.3
Individual Rate Constants and Comparison of Kinetic Models to Experimental Data

Individual rate constants for homogeneous and heterophase DADMAC homopolymerizations have been calculated from the kinetic models above. These constants are summarized in Table 4. The overall activation energies, also given in Table 4, are relatively high. These values likely result from the electrostatic repulsion between the positively charged radicals and the monomer cations.

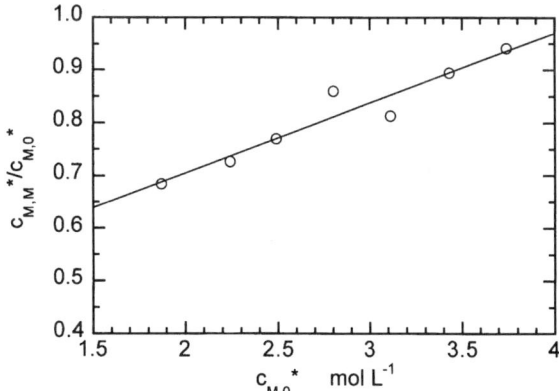

Fig. 11. Inverse-emulsion polymerization of DADMAC. Influence of the monomer concentration on the partition equilibrium of the monomer. ($c_{M,0}$*: initial monomer concentration in monomer/water droplets at equilibrium; $c_{M,M}$*: monomer concentration in micelles at equilibrium) (Data taken from [13])

Table 4. Individual rate constants and parameters for DADMAC homopolymerization

Polymerization technique	Parameter	Value	Conditions	Ref.
homogeneous in aqueous solution	k_t	$8 \cdot 10^6$ L·mol^{-1}·s^{-1}	50 °C	[51]
	k_p	90 L^2·mol^{-2}·s^{-2}	50 °C (calculated from $k_p/k_t^{0.5}$ [6])	
	k_{deg}	$1.6 \cdot 10^8$ L·mol^{-1}·s^{-1}	35 °C	[60]
	f	0.22		[6]
	$E_{A \, (overall)}$	100±5 kJ·mol^{-1}	30...50 °C	[59]
	$E_{A \, (overall)}$	64.5 kJ·mol^{-1}		
hetrogeneous inverse-emulsion	k_t	$1 \cdot 10^7$ L·mol^{-1}·s^{-1}	60.5 °C (calculated from $k_p/k_t^{0.5}$ [6])	[13]
	k_p	$1 \cdot 10^2$ L^2·mol^{-2}·s^{-2}	60.5 °C	
	$E_{A \, (overall)}$	89 kJ·mol^{-1}	47.5...65 °C	[13]
	$E_{A \, (propagation)}$	68 kJ·mol^{-1}	47.5...60 °C	

Equation (8) has been used to describe the progress of the homogeneous polymerization up to conversions of approximately 60%. Experimental and calculated conversion-time curves were in good agreement, even for the case of changing experimental conditions during the polymerization [51]. For the heterophase polymerization experimental and modeled conversion-time curves coincide well if a kinetic model based on first order initiator decomposition was applied and consideration of gel effect for conversions greater than 35% was included [13].

4.2
Copolymerization

One objective of the copolymerization of DADMAC is to obtain higher polymer molar mass since the molar mass of the homopolymer is limited by a slow chain propagation (Table 4). Therefore, DADMAC has been copolymerized with various ionogenic and nonionic monomers of which acrylamide is the most common. A variation of the general properties such as water solubility or charge density also becomes possible by copolymerization.

4.2.1
Copolymerization with Acrylamide

Although DADMAC (M_1) and AAM (M_2) were first reacted in 1959 [61], publications on the copolymerization kinetics are relatively limited. Figure 12 shows in-

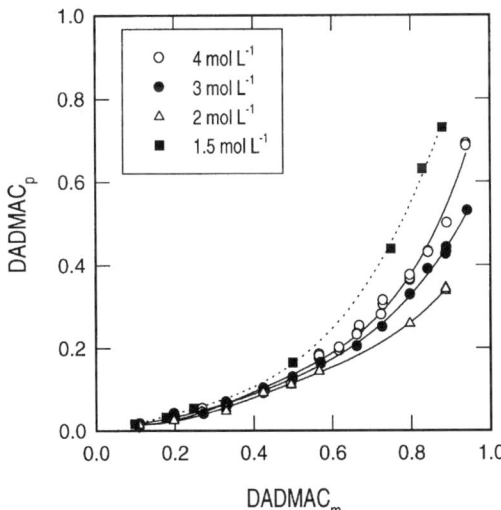

Fig. 12. Copolymerization DADMAC/acrylamide in aqueous solution. Influence of the initial monomer concentration on the copolymer composition (— data taken from [64]; ···· data taken from [65])

Table 5. DADMAC/AAM reactivity ratios

Reactivity ratios		Polymerization conditions				Analytical methods	Ref.
DADMAC r_1	AAM r_2	$[M_1]+[M_2]$ mol L^{-1}	Feed M_1 mol %	Initiator	T °C		
Solution polymerization							
0.22[a]	7.14[a]	4	20→72	APS	35	Potentiometry	[64]
0.074[a]	6.62[a]	3	11→89			Elemental analysis	
						NMR	
0.58	6.7	1.5	10→88	KPS/ isopropanol	40	Colloid titration	[65]
0.12	6.2	2		Azo	40		[66]
0.04[a]	4.8[a]	3	43→99	Azo	55	Potentiometry	[67]
0.03[a]	6.2[a]	1.5	10→99	10^{-3}–10^{-2} mol · L^{-1}		NMR	
0.03[a]	6.3[a]	0.5	10→99				
Inverse-emulsion polymerization							
0.05	7.54	5.75	20→80	Azo	47	Potentiometry	[68]
0.058	7.82	5.75	20→80	Azo	47	Colloid titration	[69]
0.172	7.28		20→80			Potentiometry	
0.06 ±0.03	6.4±0.4	0.5	30→70	Azo	50	HPLC	[70]

[a] Averaged values since the reactivity ratios were observed to depend on the feed ratios

stantaneous copolymer composition diagrams for experiments conducted under various reaction conditions.

Reactivity Ratios

In Table 5 the reactivity ratios obtained for various polymerization conditions are summarized.

All authors have obtained similar curvatures, as indicated in Fig. 12, and the following trends have been observed: $r_1 < r_2, r_1 < 1, r_2 > 1$, and $r_1 \cdot r_2 \neq 1$. Therefore, the copolymerization of DADMAC/AAM can correspondingly be classified as a nonideal nonazeotropic copolymerization. Related to the definition of the reactivity ratios [62], it follows from the experimental values in Table 5 for the individual propagation rate constants, that $k_{11} < k_{12}$ while $k_{22} > k_{21}$. Given the extremely high propagation rate constant for AAM ($k_p = 2 \cdot 10^4$ l·mol^{-1}·s^{-1} [63]), it is therefore a reasonable water soluble monomer to form a high molecular polyelectrolyte backbone. However, as Table 5 indicates, the reactivity ratios of DADMAC and AAM differ by up to two orders of magnitude, prompting some investigators to speculate that these two monomers cannot copolymerize and indeed form tenacious blends of homopolymers. This postulate has not, however been born out by spectrometric investigations (see Sect. 3).

Given the preceding mechanistic discussion, one would not expect the reactivity ratios to represent true kinetic parameters. Indeed, the reactivity ratios are sensitive to monomer and simple electrolyte concentrations, comonomer feed compositions, temperature and whether the reaction was carried out in aqueous solutions or a heterophase system. Clearly the data from the various groups are partly inconsistent, but still some general conclusions regard DADMAC/AAM copolymerization, listed in Table 6, are reasonable.

A comparison of the experimental results in Table 6 is, however, complicated by several factors:
- monomers of different purity were employed
- the copolymerizations were carried out under different experimental conditions
- the copolymers were analyzed using different methods to remove the residual monomers and to determine the copolymer composition

Table 6. Kinetic peculiarities of DADMAC/AAM copolymerization

Kinetic observation	Ref.
r_1 and r_2 increase with total monomer concentration	[64]
r_1 and r_2 vary inversely with the concentration of DADMAC in the initial comonomer solution	[64]
r_1 and r_2 are essential unchanged between homogeneous and heterogeneous polymerization process	this work
r_1 increases with ionic strength by addition of low molecular electrolyte	[69]
r_1 is insensitive to pH	[65]

- the degree of conversion is either different or unavailable
- to calculate the reactivity ratios, various models and procedures (graphical, numeric) have been applied.

With the exception of [64], the majority of copolymerizations has been carried out with non-recrystallized DADMAC. Although, there is no evidence that the monomer purity markedly influences the reactivity ratios of Table 5, a general influence on the rate of polymerization should be taken into account. The majority of analytical methods require removal of the monomers before the copolymer composition can be determined. For this reason, HPLC has been shown to provide estimates of reactivity ratios with more narrow confidence intervals [70]. Due to the differences between r_1 and r_2, particularly at higher DADMAC contents in the monomer feed, it is quite challenging to maintain a low conversion of AAM and a constant monomer feed composition.

Comparison of DADMAC with other Cationic Monomers

Comparing the reactivity ratios of the DADMAC/AAM copolymerization with results of the copolymerization of other cationic monomers with AAM, significant differences can be identified. The differences between r_1 and r_2 are much lower, and the cationic monomer even reacts preferentially during the copolymerization. As an example, for cationic methacrylic esters and methacrylamid derivatives, $1<r_1<2.5$ and $0.25<r_2<0.6$ were obtained [65, 70]. These values are relevant for a nonideal copolymerization preferring the cationic component. For the cationic analogs of acrylic acid and acrylamide, $0.34<r_1<0.48$ and $0.29<r_2<0.95$ have been published [65, 70]. These values are related to an azeotropic copolymerization, preferring the cationic monomer only at low content in the comonomer mixture.

Deviations from the Ultimate Copolymerization Model

The Mayo-Lewis model cannot completely explain the peculiarities of the DADMAC/AAM copolymerization [64]. Clearly, electrostatic interactions between the monomer cation and the charged radical chain end influence the propagation reactions because the cationic charge is situated closely to the radical position (see Sect. 4.1). For these reasons, the ionic strength, solvatation, and the charge density at the growing chain end, for the copolymerization become important. In the case of the acrylic and methacrylic compounds discussed above, the charge carrying group is situated far from the reactive double bond. It seems that this distance is large enough to screen the electrostatic repulsion between the charged monomer and the charged chain end resulting in a higher reactivity of the ionic monomer.

Application of the penultimate model to the experimental results of [64] resulted in the following reactivity ratios:

$$r_1 = \frac{k_{111}}{k_{112}} = 0.032 > r_1' = \frac{k_{211}}{k_{212}} = 0.021 \tag{10}$$

$$r_2 = \frac{k_{222}}{k_{221}} = 7.19 > r_2' = \frac{k_{122}}{k_{121}} = 2.97 \tag{11}$$

with $k_{111} < k_{112}$, $k_{211} < k_{212}$, $k_{222} > k_{221}$, and $k_{122} > k_{121}$ [38].

Assuming an influence of the charge density at the chain end on the reaction with the cationic monomer, the order $k_{111} < k_{211} < k_{121} < k_{221}$ becomes probable. With this assumption and comparing the reactivity ratios (Eqs. (10) and (11)), and from $k_{12} < k_{22}$ (Sect. 4.1), $k_{112} < k_{212} < k_{122} < k_{222}$ should be valid [38].

From these results, it can be concluded, that the structure of the growing chain end not only influences the addition of the cationic monomer but also the propagation step with acrylamide [38]. Figure 13 shows a comparison of the experimental results with the penultimate model and the best fit using the Kehlen/Tüdös equations [71]. A better agreement can be observed particularly in the range of higher DADMAC content.

The application of the "Error-in-Variables" method [72], which considers the variance in all measured parameters to obtain more precise estimates has also led to a better agreement with experimental values [73].

Copolymerization at High Conversion

Butler et al. have reported crosslinking leading to gelation for the DADMAC/AAM copolymerization at a total monomer concentration of 4 mol L^{-1} and 40 °C [66]. However, gelation only occurred for 20/80 DADMAC/AAM monomer

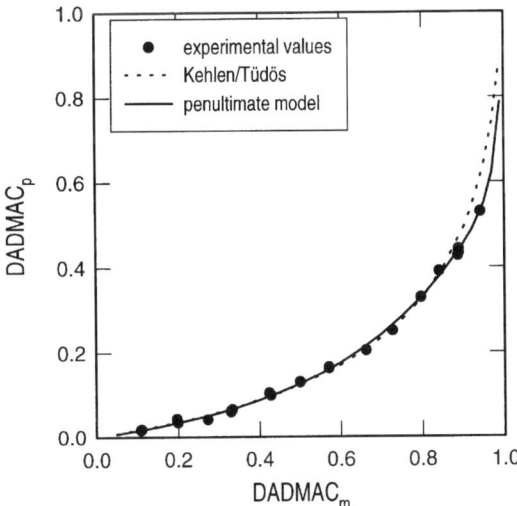

Fig. 13. Copolymerization DADMAC/acrylamide in aqueous solution. Comparison of experimental results with kinetic models (Data taken from [38])

feed, while crosslinking was observed for all monomer feed compositions. The gel point was at 51% conversion. These results have been mechanistically discussed in connection with the cyclopolymerization of DADMAC, and a significant allylic hydrogen abstraction by the growing polymer radical characteristic of allyl polymerization has been proposed [66]. Their reactivity ratios, also listed in Table 5, were in reasonable agreement with the other results. An influence of impurities from the used commercial monomer solution has not been taken into account.

Control of Copolymer Composition

Recently, a general algorithm based on a semi-batch copolymerization, with a time-dependent dosage of the more reactive monomer AAM, has been described to produce normally distributed copolymers with various charge densities but having similar molar masses [67]. Caused by the strongly divergent r_1 and r_2 values the procedure is limited to a mid range of copolymer composition and not too high conversions. For high DADMAC contents the molar mass is limited whereas for high AAM contents very high molar masses result. The influence of the copolymer composition on the molar mass has already been published in [74].

4.2.2
Copolymerization with other Monomers

A number of nonionic, anionic and cationic monomers has been used to copolymerize DADMAC in binary or multicomponent systems by various techniques. These monomers are listed in Table 7. In relation few systems have the mechanism and kinetics of the polymerization process been investigated [1]. Interest has been primarily directed toward new products or properties, with the data mostly in patents.

The free radical polymerization of DADMAC (M_1) with vinyl acetate (M_2) in methanol proceeds as a nonideal and nonazeotropic copolymerization with monomer reactivity ratios $r_1=1.95$ and $r_2=0.35$ were obtained [75]. The resulting low molar mass copolymers were reported to be water soluble over their whole range of composition. Modification of the vinyl acetate unit by hydrolysis, acetalization, and acylation resulted in DADMAC products with changed hydrophilic or polyelectrolyte properties [75]. For the copolymerization of DADMAC and N-methyl-N-vinylacetamide (NMVA) a nearly ideal copolymerization behavior could be identified [45]. The application properties of the various copolymer products will be discussed in Sect. 8.

Table 7. Summary of monomers copolymerized with DADMAC

Neutral comonomers	Anionic comonomers	Cationic comonomers
Acrylnitrile [76]	Acrylic acid [82, 87–89]	Diallyldecylmethylammonium chloride [90]
Acrolein [77]	2-Acrylamido-2-methylpropane sulfonic acid [83, 88]	Diallylmethyl(1,1-dihydropentadecafluoro-octoxyethyl)ammonium chloride [90]
N-isopropylacrylamide [78]	Methacrylic acid [89]	Diallyldecylammonium chloride [91]
N-metyl-N-vinylacetamide [45]	Sodium styrene sulfonate [88]	Diallylhaxadecylammonium chloride [91]
N-vinylformamide [79]		Diallyldecylmethylammonium chloride [92]
Styrene [80]		Diallylhexadecylmethylammonium chloride [92]
Sulfur dioxide [81]		Diallylguanidine acetate [93]
Vinylacetate [82]		Methacryloxyethyltrimethylammonium chloride [94]
Vinylalcohol [83]		
Vinyltrialkylsilane [84–86]		

5
Solution Properties of Diallyldimethylammonium Chloride Polymers

5.1
Fundamentals

The study of several polyelectrolytes through various experimental methods has led to diverging results and controversial conclusions. This situation has recently been summarized from the theoretical point of view in [28, 95–97]. However, there are some interesting data resulting from experiments with PDADMAC and DADMAC copolymers which remain unexplained. The aim of this Section is to present these experimental results and, furthermore, to discuss the data in terms of existing polyelectrolyte theories. For a better understanding of the experimental results under discussion a short fundamental summary of the main properties and parameters shall be given. However, it is not the aim of this review to evaluate the various theoretical approaches.

5.1.1
Polyelectrolyte Models and Counterion Condensation

Counterions are necessary to ensure electroneutrality in polyelectrolyte solutions. Therefore, it can be energetically advantageous if a fraction of counterions are situated in the vicinity, or at the surface, of the polyion in order to reduce the charge of the polyion. To answer the question under which conditions this occurs, the concept of the counterion condensation has been introduced by Fuoss, Katchalsky and Lifson [98], Alexandrowicz and Katchalsky [99] or Oosawa [100] and subsequently theoretically developed by Manning [101–108].

A linear charge density parameter (Manning parameter) has been defined as:

$$\xi = \frac{l_B}{b} \tag{12}$$

where b is the spacing between singly charged groups along a infinitely long linear polyelectrolyte chain, calculated as b=L/N with L the contour length and N the number of charged groups. l_B is the Bjerrum length, which has a value of 0.712 nm in water at 20 °C. The theory predicts a maximum of the effective charge density. For instance, if we have monovalent counterions it must be $\xi_{eff} \leq 1$. Thus, counterion condensation occurs if b is smaller than the Bjerrum length. The fraction of condensed counterions is equal to $1-\frac{1}{\xi}$. Hence, the Bjerrum length appears to be the smallest distance between elementary charges. The fraction of uncondensed ions is then $\frac{1}{\xi}$.

5.1.2
Concentration Regimes

In pure polyelectrolyte solutions a decreasing polyelectrolyte concentration c_p is followed by an increase of the Debye length l_D and an increase in chain stiffness. Applying scaling concepts [109] and considering an electrostatic contribution to the persistence length L_p [110–113] various concentration regimes could be identified for polyelectrolyte solutions. Odijk derived different critical concentrations [111]:

$$c^* \approx \left(N_A L^2 a\right)^{-1} \qquad L_p \gg L \tag{13}$$

$$c_b^* \approx \left(16\pi N_A L a b\, \xi\right)^{-1} \qquad L_p \approx L \tag{14}$$

$$c^{**} \approx \left(32\pi^2 N_A a b^2 \xi^2\right)^{-1} \qquad L_p < L \tag{15}$$

where a is the monomer length and N_A is the Avogadro's number. The concentration regimes are characterized by
- highly diluted $c_p < c^*$
- transition $c^* < c_p < c_b^*$ and $c_b^* < c_p < c^{**}$
- semi-diluted $c^{**} < c_p$

The influence of the degree of polymerization and contour length respectively, on the critical concentrations are given in Table 8 for PDADMAC.

If we take into consideration that the lowest experimentally possible polyelectrolyte concentration c_p is approximately 10^{-6} monomol L^{-1}, it follows from Table 8 that the diluted solution state, $c_p < c^*$, cannot be realized for PDADMAC if N >2000, i.e. if M_n >320,000 g·mol^{-1}. The theoretical treatment and the experimental studies of the concentration dependent behavior of polyelectrolytes in solution is usually restricted to the case with or without an excess of a low molecular electrolyte. A relatively limited amount of data exist for similar concentrations of polyelectrolytes and low molecular mass salt [97].

Table 8. Influence of the degree of polymerization (N) on the critical concentrations of DADMAC in aqueous solution (a=0.5 nm, b=0.5 nm, T=20 °C)[1]

Degree of polymerization N	Contour length [nm] L	Critical concentrations [monomol l^{-1}]		
		c^*	c_b^*	c^{**}
50	25	5.32·10^{-3}	3.71·10^{-3}	2.08·10^{-2}
100	50	1.33·10^{-3}	1.86·10^{-3}	"
500	250	5.32·10^{-5}	3.71·10^{-4}	"
1000	500	1.33·10^{-5}	1.86·10^{-4}	"
5000	2500	5.32·10^{-7}	3.71·10^{-5}	"

[1] a: monomer unit length, b: charge distance

5.1.3
Interactions Between Polyion and Counterions

Different methods are suitable to investigate interactions between the polyion and the counterions. Dependent on the method employed, various parameters can be determined, which describe these interactions. These include:
- the osmotic coefficient f_o
- the activity coefficient of the counterions f_a
- the self diffusion parameter D_i / D_i^o

For $\xi = \dfrac{l_B}{b} > 1$ Manning has derived the correlation [104]

$$f_o < f_a = 1.21 f_o < \frac{D_i}{D_i^o} = 1.74 f_o = 1.438 f_a \tag{16}$$

The theoretical relationships for f_a and D_i/D_i^o from Manning's theory [102–108] as well as from other theories, for example, of Iwasa [114], Gueron [115], and Yoshida [116, 117] have been summarized in Table 9.

5.1.4
Electrolytic Conductivity

The basic equation for the equivalent conductivity of a pure polyelectrolyte solution is [118–122]:

$$\Lambda = f_c \left(\lambda_p + \lambda_c^o \right) \tag{17}$$

where Λ is the equivalent conductivity of the solution, λ_c^o is the equivalent conductance of the counterion in an infinitely diluted solution without polyions, and λ_p is the equivalent conductance of the polyion [123].

The parameter f_c is given by [119, 123] as:

$$f_c = \frac{D_i}{D_i^o} \tag{18}$$

In the model, λ_p is determined by a temperature dependent electrophoretic mobility factor [123] which contains the viscosity of the solvent as well as its relative permittivity, λ_c^o, the radius of the polymer chain and the Debye screening length l_D. The following equation holds for the case that electrolyte and polyelectrolyte are in the same concentration range:

$$l_D = \left[4\pi N_A l_B \left(\xi^{-1} c_p + 2 c_s \right) \right]^{-\frac{1}{2}} \tag{19}$$

It can be simplified for polyelectrolyte solutions in the absence of a low molecular weight electrolyte [27]. Note that l_D has been discussed from different points of view [124]. Here we provide only the generally accepted definition [27].

Diallyldimethylammonium Chloride and its Polymers

Table 9. Equilibrium and transport coefficients. ($X = c_p / c_s$)

Coeff.	$\xi < 1$	$\xi > 1$
Coefficients in aqueous solution		
	Manning≡Iwasa	
f_a	$\exp\left(-\dfrac{\xi}{2}\right)$	$\dfrac{1}{\xi}\exp\left(-\dfrac{1}{2}\right)$
	Gueron	
f_a	–	$0.7\xi^{-1}$
	Manning	
D_i/D_i°	$1 - 0.55\xi^2(\xi + \pi)^{-1}$	$0.87\xi^{-1}$
	Yoshida	
D_i/D_i°	–	$0.33 + 0.43\xi^{-1}$
Coefficients in aqueous solution with addition of salt		
	Manning	
f_a	$\exp\left(-\dfrac{\xi X}{2X+4}\right)$	$\dfrac{\dfrac{X}{\xi}+1}{X+1}\exp\left(\dfrac{-\dfrac{X}{2\xi}}{\dfrac{X}{\xi}+2}\right)$
	Iwasa	
f_a	$\exp\left\{\dfrac{\xi X}{X+2}\left[-\dfrac{1}{2} + 0.39\xi\left(\dfrac{X}{X+2}-1\right)\right]\right\}$	$\dfrac{\dfrac{X}{\xi}+1}{X+1}\exp\left\{\dfrac{\dfrac{X}{\xi}}{\dfrac{X}{\xi}+2}\left[-\dfrac{1}{2} + 0.39\left(\dfrac{\dfrac{X}{\xi}}{\dfrac{X}{\xi}+2}-1\right)\right]\right\}$
	Gueron	
f_a	–	$\left(0.7\dfrac{X}{\xi}+1\right)(X+1)^{-1}$
	Manning	
D_i/D_i°	$1 - \dfrac{1}{3}\left[\dfrac{X\xi}{2+X\left(1+\dfrac{\pi}{\xi}\right)}\right]$	$\dfrac{\dfrac{X}{\xi}+1}{X+1}\left\{1 - \dfrac{X}{3\xi}\left[\dfrac{X}{\xi}(1+\pi)+2\right]^{-1}\right\}$

5.2
Experimental Results

5.2.1
Counterion Activity

The counterion activity of DADMAC polymers has been determined by direct potentiometry using chloride ion selective electrodes. If the polyelectrolyte concentration of a solution is known the counterion activity coefficient (f_a) can be calculated from

$$f_a = \frac{a_c}{c_p} \tag{20}$$

where a_c is the measured counterion activity of the polyelectrolyte solution [125].

Effect of Molar Mass

Figure 14 shows the experimental counterion activity coefficients for DADMAC, PDADMAC of different molar masses M_n (with $M_w/M_n \approx 1.5$), as well as the theoretical concentration dependence for a charge distance of b=0.5 nm, calculated in terms of different theoretical expressions from Table 9.

The activity coefficients of the polymers are much lower than those of the monomer. Theories predict, within their limits, a concentration, molar mass, and chemical structure independence of the counterion activity. As seen from Fig. 14, the experimental curvatures differ from the theoretical predictions. The concentration dependence and the absolute values of f_a change with the molar mass. Further, the activity coefficient has been found to be reciprocally related to the molar mass [38]. To obtain reliable results a minimization of the salt out-

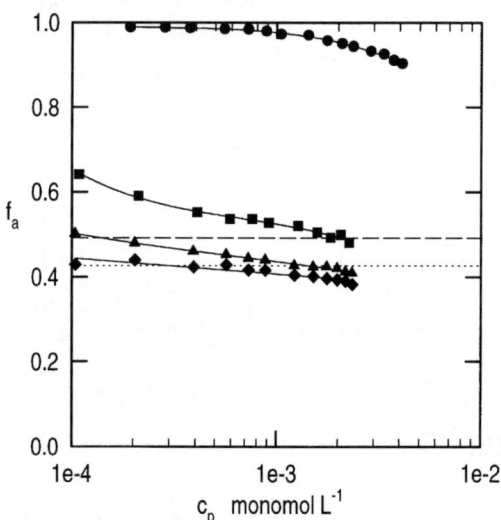

Fig. 14. Concentration dependence of the counterion activity coefficients, f_a, of DADMAC, PDADMAC with different molar masses, and comparison with theoretical predictions (T=20 °C; ● DADMAC; PDADMAC with M_n: ■ 12,000 g mol^{-1}; ▲ 72,000 g mol^{-1}; ♦ 170,000 g mol^{-1}; ······Manning; – – – Gueron) (Data taken from [38])

flow from the reference electrode, controlled by electrical conductivity measurements, must be realized [38]. This salt outflow from the reference electrode may be responsible for the higher values of counterion activity for PDADMAC published by other authors [126].

Effect of Charge Density

According to theoretical considerations, f_a increases if the charge distance on the polyelectrolyte chain becomes larger. This could be shown for DADMAC/AAM copolymers for charge distances in the range of 0.5 to 1.9 nm (Table 10). For $b \leq l_B$ the f_a values are within the limits of the Manning theory and below the values calculated by Gueron. However, for $b \geq l_B$ the experimental values are remarkably below the theoretical values (Table 10).

Effect of Chemical Structure and Chain Architecture

The agreement between theory and experimental results has been found to depend on the chemical structure of the repeat unit. Figure 15 represents the concentration dependence of counterion activity coefficients for two polyelectrolytes, PDADMAC and MEAC/AA (50:50) copolymer, differing in chemical structure of the cationic monomer unit but having the same charge distance (0.5 nm).

In terms of Manning's theory a comparison of experimental f_a with theoretical calculations leads to a better agreement for PDADMAC having the ammonium group near the polymer backbone. For larger spacer groups between the polymer backbone and the ionic group the counterion activity was found to be higher than predicted for a line charge. An increase of the counterion activity has also been identified for branched PDADMAC, and explained by a large number of end groups along the polymer chain [38].

Table 10. Experimental and calculated counterion activity coefficients (f_a) for DADMAC/AAM copolymers with different charge distances (b)

DADMAC mol %	b nm	f_a experimental[a]	f_a calculated Manning	Gueron
15	1.92	0.77 … 0.73	0.84	---
38	0.91	0.61 … 0.58	0.69	---
41	0.86	0.60 … 0.53	0.67	---
56	0.70	0.60 … 0.48	0.60	0.69
69	0.61	0.57 … 0.46	0.53	0.60
100	0.50	0.42 … 0.39	0.43	0.49

[a] experimental range: $c_p = 10^{-4} - 2 \cdot 10^{-3}$ monomol L^{-1}

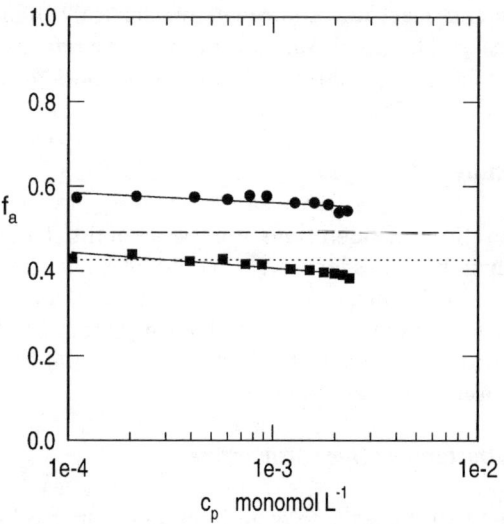

Fig. 15. Influence of the chemical structure of the ionic monomer unit on the concentration dependence of the counterion activity coefficients f_a; charge distance b= 0.5 nm (T=20 °C; ● copolymer of acrylamide with 50 mol % 2-(methacyloyloxy)ethyltrimethylammonium chloride, M_w=8.6 ·10^6 g mol^{-1}; ■ PDADMAC, M_n=1.7·10^5 g mol^{-1}; · · · · · Manning; - - - Gueron) (Data taken from [38])

Influence of Low Molecular Mass Electrolyte

The influence of molar mass, charge density as well as chain branching was also determined in the presence of low molecular mass salt. As seen in Fig. 16, the differences between theory and experiment are more important to low molar masses. In Fig. 16 the concentration dependence of the activity of the low molecular salt has been taken into account when calculating $f_a^c = f_{exp}/f_0$ [114, 126], where f_a^c and f_{exp} are calculated and experimentally determined counterion activity coefficients, respectively; f_0 is the activity coefficient of the added low molecular salt in aqueous solution without polyelectrolyte.

5.2.2
Electrolytic Conductivity

Conductivity investigations have been performed by measuring the specific conductance as function of c_p, the equivalent concentration of polyelectrolyte. According to [128] the specific conductance can be expressed as

$$\kappa = \kappa_o + \Lambda^o c_p + c_p \Phi(c_p) \tag{21}$$

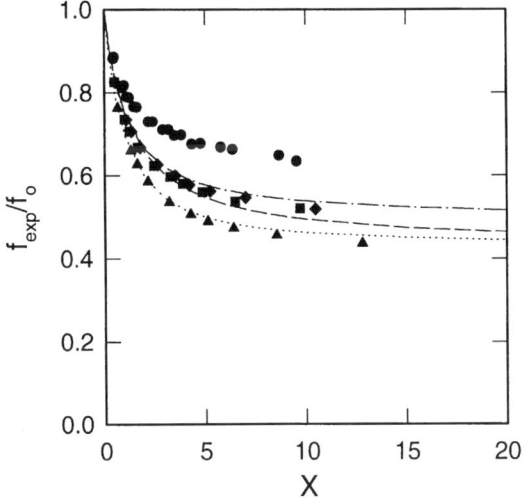

Fig. 16. Influence of the ionic strength on the counterion activity of PDADMAC with different molar masses. Variation of the ionic strength by addition of NaCl. ($X=c_p / c_s$; $c_p=10^{-3}$ monomol L^{-1}; T=20 °C; PDADMAC M_n : ● 12,000 g mol^{-1}; ◆ 22,000 g mol^{-1}; ■ 72,000 g mol^{-1}; ▲ 170,000 g mol^{-1}; – · – · – Gueron; – – – Manning; · · · · · Iwasa) (Data taken from[38])

where Λ^o is the equivalent conductivity at infinite dilution, κ_o reflects the conductivity of the solvent and the function $\Phi(c_p)$ represents the effect of interionic interaction on conductivity. From the linear initial range of κ vs. c_p plot, the extrapolated value of the solvent conductivity (κ_o'), and Λ^o can be determined. For reliable Λ^o, a good agreement between κ_o and κ_o', the directly measured and the extrapolated value, is necessary.

Effect of Concentration

Figure 17 shows the concentration dependence of κ for PDADMAC of different molar masses. The curvature increases slightly with dilution.

The plot of the equivalent conductivity vs. the polyelectrolyte concentration, however, is more suitable to demonstrate the concentration dependent changes of the polyelectrolyte conductivity [129]. Generally, the equivalent conductivity (Eq. (17)) can be written as [128]

$$\Lambda = \frac{\kappa - \kappa_o}{c_p} = \Lambda^o + \Phi(c_p) \tag{22}$$

The concentration dependence of this equivalent conductivity is given in Fig. 18 for the low molecular salt NaCl, the monomer DADMAC, and PDADMAC.

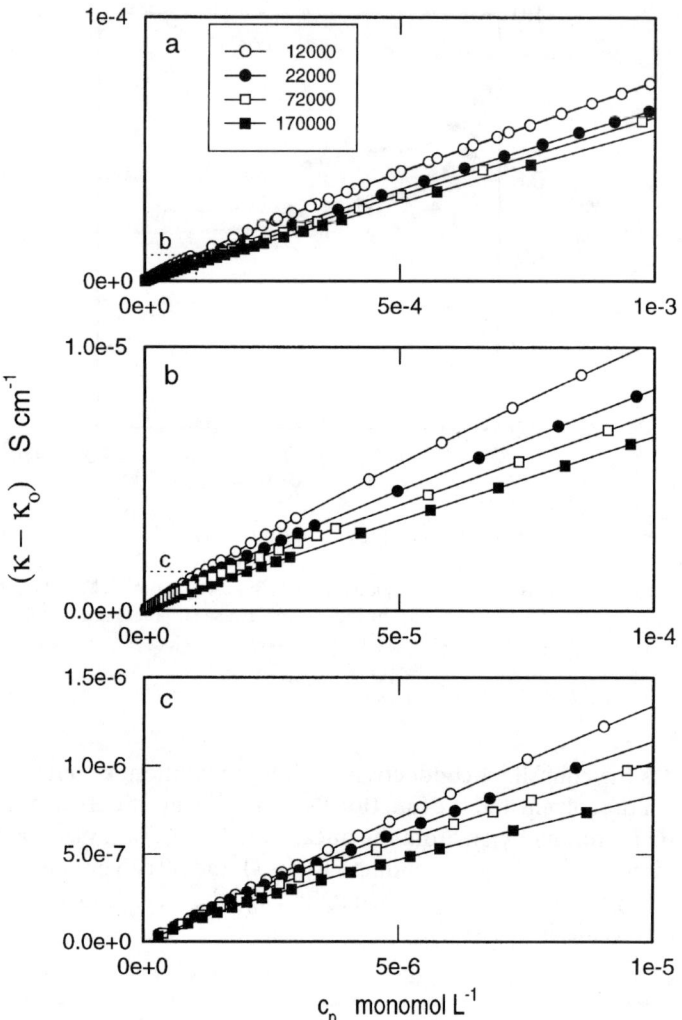

Fig. 17. Concentration dependence of the specific conductance of PDADMAC with different average molar masses M_n (T=20 °C) (Data taken from [38])

The curvature of the monomer solution is similar to that of the salt. The concentration dependence between PDADMAC samples are significantly different.

Effect of Molar Mass

Assuming infinitely long polyelectrolyte chains, the Manning theory neglects the influence of molar mass on the equivalent conductivity. It could be shown in

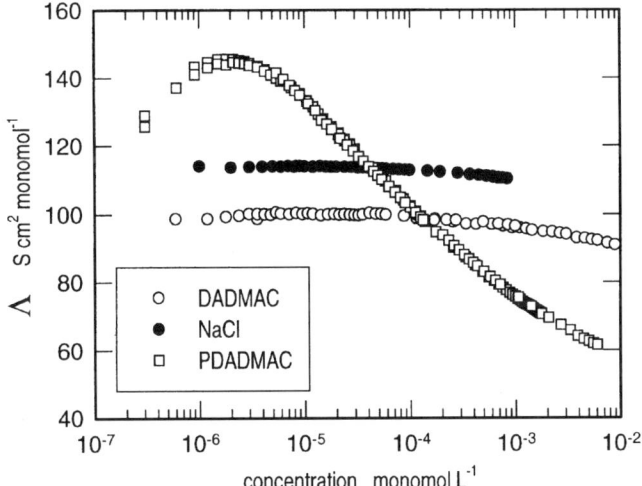

Fig. 18. Concentration dependence of the equivalent conductivity Λ for low molecular electrolytes (NaCl, DADMAC) and PDADMAC. PDADMAC: M_n=12,000 g mol^{-1}; T=20 °C (Data taken from [38])

[130] that the equivalent conductivity of PDADMAC is strongly influenced by the molar mass, particularly, at high dilutions. Further, a correlation between the strong increase of Λ and the concentration regimes seems to exist. Λ strongly increases below the overlap concentration c* [38, 130]. A more detailed study of the concentration dependence of the equivalent conductivity in the presence of low molecular salt has led to correlation $\Lambda_{max} \sim c_s^{-0.18 \pm 0.02}$ and $\Lambda_{max} \sim M_n^{-0.17 \pm 0.02}$ [38, 130].

Effect of Ionic Strength

The influence of NaCl on the equivalent conductivity of PDADMAC with different molar masses is demonstrated in Figs. 19 a and b. Λ_{max} decreases with increasing c_s and is observed at higher c_p.

Plotting Λ vs. the ratio of the polyelectrolyte to the salt concentration, c_p/c_s, the largest change of the slope is located in the c_p/c_s region between 1 and 3. An example is given in Fig. 20 for the lowest molar mass and holds for all ionic strengths and molar masses that have been investigated. This implies that a linear increase of the equivalent conductivity below the overlap concentration will only be found if the polyelectrolyte concentration exceeds the concentration of monovalent low molecular electrolyte by a factor of two to three.

A satisfactory explanation of the experimental results in terms of the existing theories cannot be provided. An interpretation using a scaling of the activity coefficient vs. the ratio of the Debye length l_D to the contour length L has been discussed in [38] for narrow distributed NaPSS standards. This interpretation,

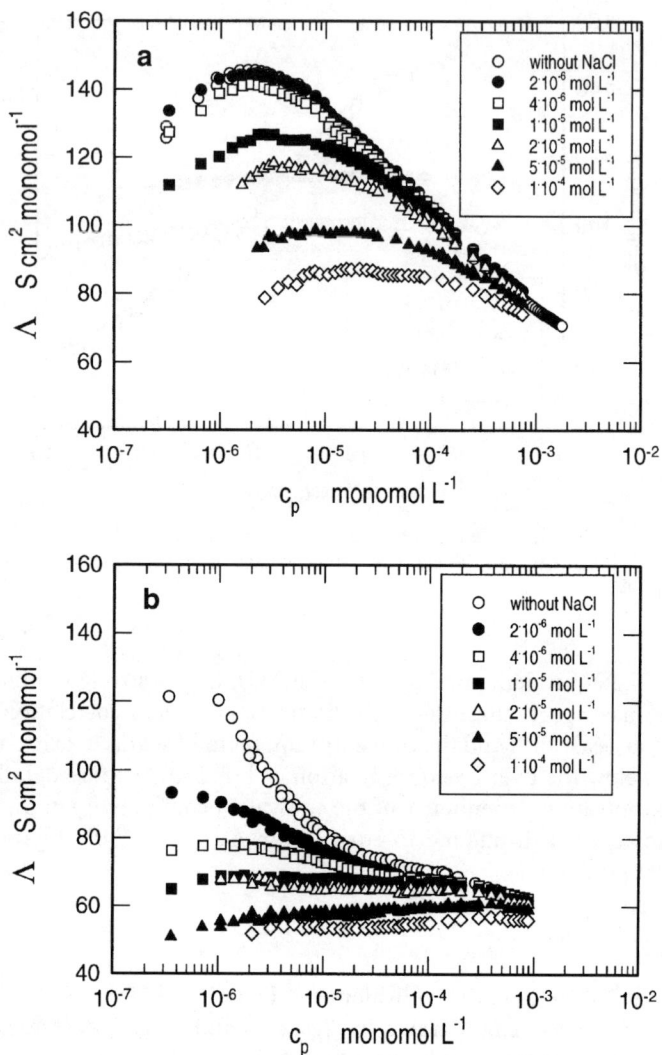

Fig. 19 a, b. Concentration dependence of the equivalent conductivity Λ. Influence of the ionic strength c_s for PDADMAC. **a:** M_n=12,000 g mol^{-1}, **b:** M_n=325,000 g mol^{-1}; T=20 °C (Data taken from [38])

based on an increase of f_c as the result of the decreasing Coulombic interaction in the case of $l_D > L$, also holds for PDADMAC. The scaling of $f_{c,fit}$ as a function of l_D/L has yield $f_{c,fit}=0.612(l_D/L)^{0.25}$ for M_n=12,000 g·mol^{-1} and $f_{c,fit}=0.612(l_D/L)^{0.26}$ for M_n=22,000 g·mol^{-1}. For the higher molar masses the scaling has been influenced by the increasing polydispersity of the samples [38]. Using

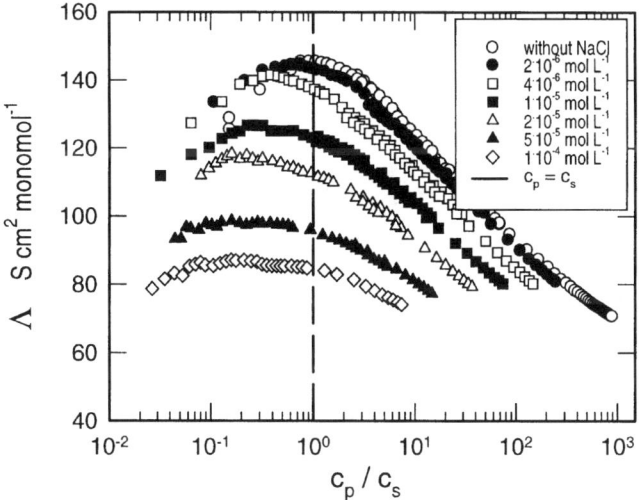

Fig. 20. Plot of the equivalent conductivity Λ vs. the ratio of the PDADMAC to the salt concentration c_p / c_s for PDADMAC with M_n=12,000 g mol^{-1}; T=20 °C

average values, the molar mass and ionic strength dependence of the equivalent conductivity in the highly diluted solution is given by [38]:

$$\Lambda = 0.61(l_D / L)^{0.25}\left(\lambda_p^o + \lambda_c^o\right) \qquad (24)$$

Effect of Charge Density

The influence of the charge density on the electrolytic conductivity is demonstrated for high molar mass PDADMAC and the AAM-copolymers in Fig. 21.

Changing the charge distance from 0.5 up to 1.92 nm the conductivity behavior changes remarkably. For $b > l_B$, a weak and nearly linear increase of Λ has been found with increasing dilution. However, for b smaller than or in the range of l_B, Λ increases in the same manner as described above for the different molar masses of the homopolymer.

Effect of Temperature

The influence of the temperature in the range 20 °C <T <50 °C on the conductivity behavior of PDADMAC is demonstrated for two molar masses in Fig. 22 a and b respectively.

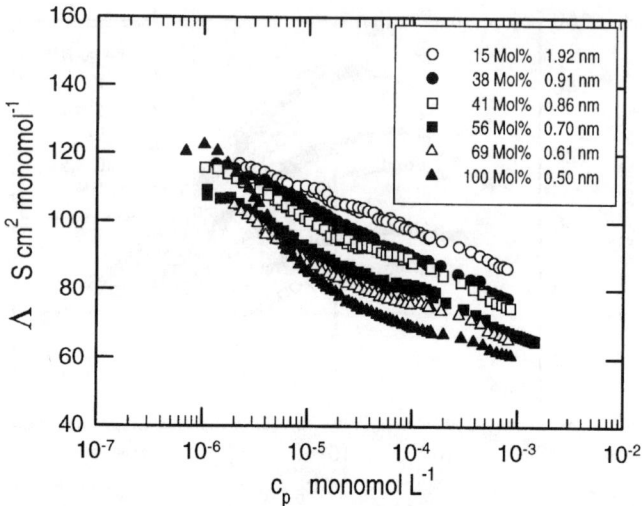

Fig. 21. Influence of the charge density on the concentration dependence of the equivalent conductivity Λ (T=20 °C). The indicated charge distance b was realized by copolymerization of DADMAC with AAM. The percentage denotes the DADMAC content in the copolymer (Data taken from [38])

A shifting of Λ_{max} could not be identified. Additionally, Fig. 23 shows the temperature dependence of the limiting values Λ_{max} for DADMAC and PDADMAC with different molar masses. The two lower molar mass samples are more strongly influenced by the temperature than the high molar mass polymer and the monomer. However, to compare the different molar masses the viscosity of the solutions has to be considered (Walden law [39]). In order to finalize the conclusions additional experiments will be necessary [38].

As is evident from Fig. 23, for virtually all molar masses and temperatures, Λ_{max} of the polymers was found to be higher than the values for the monomer. Different hydration of the isolated monomer ion and the monomer units in the polymer chain is assumed to be responsible for these findings [38].

Although only a limited amount of experimental data is available which describes the interaction between the polyion and the counterions for PDADMAC compounds, the influence of macromolecular and structural parameters on the counterion activity and the electrolytic conductivity in semidiluted, and particularly in highly diluted aqueous solutions, can be clearly shown. All these findings are not only of scientific interest but also of practical relevance. Since for many technical applications PDADMAC concentrations in the ppm range are usual, this basic knowledge of the molecular and medium influences gives more insight into the action mechanism of these polyelectrolytes and may assist in the optimization of several applied processes.

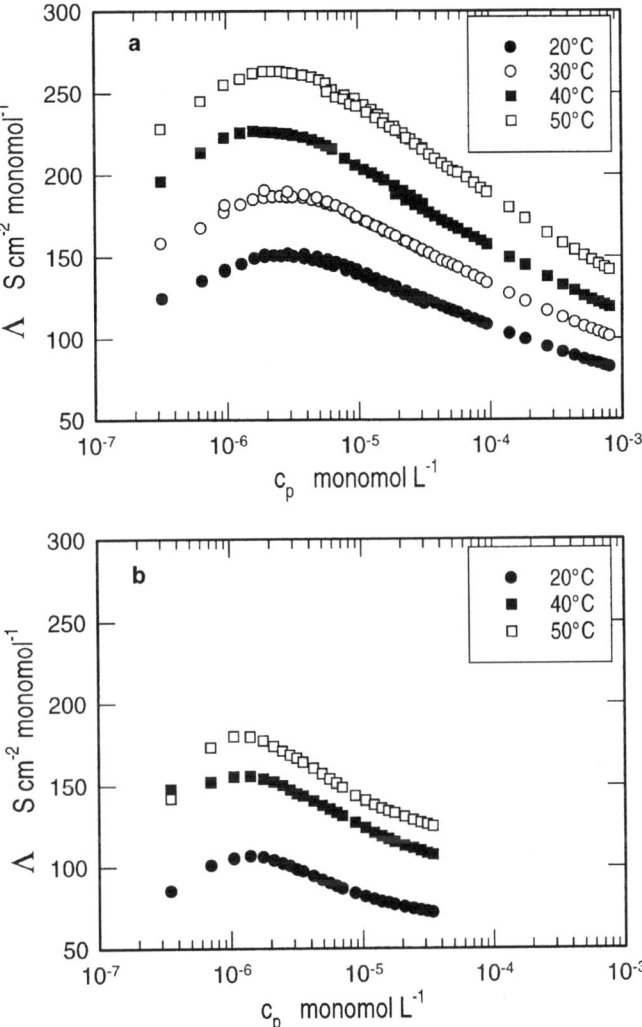

Fig. 22 a, b. Influence of the temperature on the concentration dependence of the equivalent conductivity Λ for PDADMAC with different molar masses. **a:** M_n=12,000 g mol^{-1}; **b:** M_n= 170,000 g mol^{-1} (Data taken from [38])

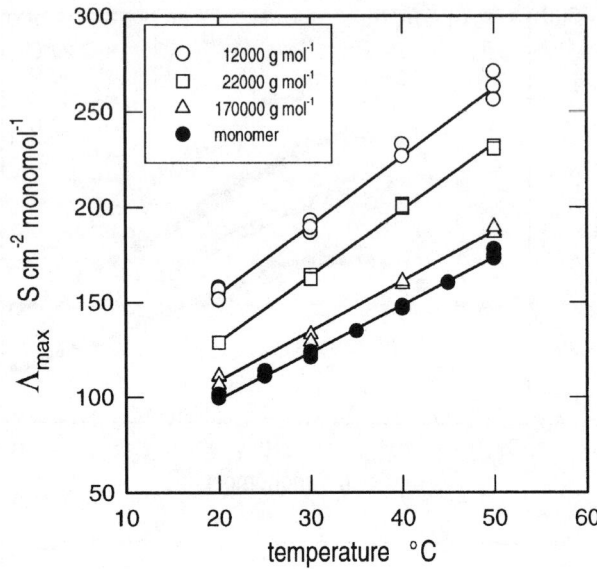

Fig. 23. Temperature dependence of the maximum equivalent conductivity Λ_{max} for PDADMAC with different molar masses and DADMAC (Data taken from [38])

6
Molecular Characterization of Diallyldimethylammonium Chloride Polymers

6.1
General Considerations

The molecular characterization of polyelectrolytes in general, and of DADMAC polymers in particular is complicated for several reasons. First, in aqueous solution the individual properties of the macromolecules are dominated by Coulombic interactions. Therefore, the resulting polyelectrolyte effects have to be suppressed through the addition of low molecular electrolyte, such as NaCl. The increase of the ionic strength results in a decrease of the chain stiffness of the polyelectrolyte molecules (see Sect. 5). The chains then revert to the coil dimensions of neutral macromolecules in dilute solutions. However, problems may still arise, particularly since the mode of action of these effects is quite different in various characterization methods [27].

Secondly, DADMAC polymers are heterogeneous substances. They posses distributions with respect to:
- molar mass (polydispersity, see Sect. 4)
- molecular architecture (linear, branched, crosslinked, see Sect. 3)
- chemical composition (copolymers, see Sect. 4.2)
- aggregation and self-association

Aggregation or association strongly disturb molecular characterization [27, 38]. As discussed in the previous sections the action of the low molecular electrolyte has been found to be different for PDADMACs of various molar masses. Therefore, the ionic strength must be optimized for each given system. In some cases, particularly at high polydispersities, there may not exist a low molecular salt concentration at which the ionic strength is sufficient to suppress the polyelectrolyte effects but low enough to avoid aggregation of the largest macromolecules. This is a major problem. This type of aggregation has been proposed to be treated as a self-associated process which depends on both salt and PDADMAC concentration and cannot easily be removed by filtration or centrifugation [130]. This heterogeneity has to be distinguished from permanent aggregates or gel-particles resulting either from chemical crosslinking during the polymerization process or from the existence of hydrophobic domains, for example in copolymers. These latter are often removable by filtration or centrifugation procedures. Generally, sample preparation must be optimized and carried out very carefully [27, 38, 74] to avoid aggregation and to obtain homogeneous highly diluted solutions.

6.2
Determination of Molar Masses

6.2.1
Homopolymers

Several authors have published the method for determining molar masses of DADMAC polymers, primarily in connection with practical applications [1]. In Table 11 intrinsic viscosity-molar mass relations of PDADMAC are summarized in the form of the Mark-Kuhn-Houwink-Sakurada (MKHS) relationship. The relatively high exponent of the relationships is attributed to the greater chain stiffness in comparison with vinyl backbones. One has to look quite skeptically at the values from reference [59] given its deviation from the remainder of the published data.

Dubin et al. [135] have studied the molar mass characterization by light scattering with commercial PDADMAC products "Merquat 100" and "Catfloc". The determined values for M_w in 0.4 mol·L^{-1} NaCl solution were $(2.8\pm0.8)10^5$ and $(4.5\pm3.0)10^6$ g·mol, respectively. Erratic and sometimes abrupt downward curvature was observed by a Zimm plot at scattering angles $\theta<45°$. This was attributed to the dust from these samples. Following the fractionation of the Merquat 100 sample by the preparative size-exclusion chromatography, linearity of the MKHS in 0.5 mol·L^{-1} NaCl solution was identified only for fractions with M_w $<2\cdot10^5$ g·mol^{-1} [136]. At higher M_w curvature appeared in the molar mass dependencies which was explained by the presence of chain branching [136]. Similar results were reported by Görnitz et al. [131] investigating unfractionated laboratory synthesized DADMAC polymers. Only for samples having M_w less than 10^5 g·mol^{-1} exhibited "normal" solution behavior and agreement between re-

Table 11. Exponents and constants of the Kuhn-Mark-Houwink-Sakurada relationship $[\eta]=KM^a$ for PDADMAC in 1 mol L^{-1} NaCl ($[\eta]$ in cm$^3 \cdot$g^{-1}, M in g·mol^{-1})

Molar mass	K	a	Method	Experimental conditions	Ref.
M_w	$4.61 \cdot 10^{-3}$	0.81	LS	T=25 °C, pH=5.5 $8.9 \cdot 10^4 < M_w < 4.7 \cdot 10^5$	[132]
M_w	$4.25 \cdot 10^{-3}$	0.80	LS	T=25 °C $8.6 \cdot 10^4 < M_w < 3.6 \cdot 10^5$	[74]
M_w	$4.16 \cdot 10^{-2}$	0.68	UC	T=30 °C $1.8 \cdot 10^4 < M_w < 5.4 \cdot 10^4$ (fractionated samples)	[134]
M_w	$4.83 \cdot 10^{-3}$	0.88	UC	T=30 °C $2.4 \cdot 10^4 < M_w < 1.4 \cdot 10^5$	[134]
M_n	$1.12 \cdot 10^{-2}$	0.82	OS	T=26 °C $1.6 \cdot 10^4 < M_n < 7.4 \cdot 10^4$	[133]
M_w	$1.26 \cdot 10^{-1}$	0.51	?	T=26 °C $1.3 \cdot 10^4 < M_w < 2.25 \cdot 10^4$	[59]

sults from light scattering and analytical ultracentrifugation measurement was found to exist [131]. On the contrary, for low conversion PDADMAC samples polymerized in Jaeger's laboratory, linear $[\eta]$ – M and R_g -M relations were obtained, even for high molar masses [137]. The second virial coefficients found by sedimentation equilibrium studies are higher ($1-2.5 \cdot 10^{-3}$ ml·mol·g^{-2} [131]) than those reported for light scattering measurements ($2.53 \cdot 10^{-4}$ ml·mol·g^{-2} [135] and $4 \cdot 10^{-4}$ ml·mol·g^{-2} [74]).

The nonlinearity of Zimm diagrams at small scattering angels was also reported by Dautzenberg et al. [138] and was attributed to the fact that, in particular, the technical products contain particles or associates as described in Sect. 3 of this paper. Small amounts of such particles can produce higher apparent molar masses by light scattering. The interpretation of the scattering curves by a bimodal system leads, after a theoretical curve separation, to a linearization and gives the correct molecular parameters of PDADMAC [138]. Bekturov [139] reported on molar mass determination by means of an analytical ultracentrifuge using methanol as a solvent.

6.2.2
Copolymers

In the case of DADMAC copolymers, the charge density and it's distribution must be taken into consideration. For DADMAC/AAM the copolymer composition differs strongly from the initial monomer ratio and changes depending on the conversion of polymerization due to the very different reactivity ratios r_1 and r_2 (see Sect. 5.2). Depending on the content of DADMAC units the polyelectrolyte character is more or less developed. Generally, a strong increase of the molar mass with decreasing DADMAC content has been reported [74].

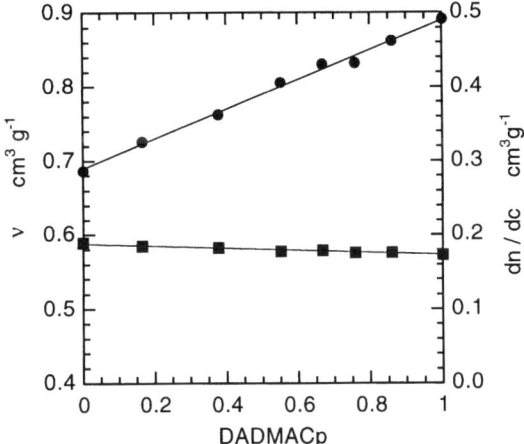

Fig. 24. Dependence of the partial specific volume v̄ (●) and refractive index increment dn/dc (■) on the DADMAC weight fraction in DADMAC/AAM copolymers (0.5N NaCl; T=20 °C) (Data taken from [46])

For the analysis of light scattering experiments the refractive indices of the DADMAC/AAM solutions at dialysis equilibrium were determined, showing the validity of the additivity principle [67]. The additivity could also be proven for the partial specific volumes which are necessary to calculate molar masses from ultracentrifugation experiments [131]. These dependencies are summarized in Fig. 24.

The second virial coefficient A_2 from osmometry, light scattering, and sedimentation experiments were found in the same range, $4-10 \cdot 10^{-7}$ mol·l·g^{-2} for DADMAC contents 8–100 mol %. The change is mainly influenced by the changes in charge density [46, 67, 131].

Correlations

R_g-M and [η]-M relations could be established for the practical use at high ionic strength (0.5 m NaCl solution), despite the various chemical compositions of the DADMAC/AAM copolymers [46, 67]:

$$R_g \text{ (nm)} = 0.117 \, M_w^{0.47} \tag{22}$$

$$[\eta] \text{ (ml/g)} = 5.34 \cdot 10^{-3} \, M_n^{0.89} \tag{23}$$

$$[\eta] \text{ (ml/g)} = 0.638 \, M_w^{0.47} \tag{24}$$

Molar masses from light scattering experiments reported by other authors [140] seem to be too high in comparison with the results above.

6.3
Determination of Molar Mass Distributions

6.3.1
Chromatography

Size exclusion chromatography (SEC) has been applied to investigate the molar mass distribution of DADMAC polymers. Generally, it must be mentioned that the exclusion effect is often superimposed by electrostatic adsorption because ionic groups are present in most aqueous chromatographic packings [141]. Since the first experiments were performed to separate cationic polyelectrolytes by means of SEC, several packing materials, calibration methods, and mobile phases have been evaluated for cationics. The development of the investigations has been described in [142] and relevant practical information on packing materials, soluents and calibration is given in [143]. Table 12 contains a summary of packing materials and experimental conditions applied to DADMAC and DADMAC copolymers.

All investigators emphasize the importance to check the packing materials for every special cationic polymer since the cationic charge and the chemical structure of the monomer units influence the chromatographic separation. Calibration has been difficult in such cases where the polydispersity of standards and

Table 12. Size exclusion chromatography of DADMAC polymers

Polymer	Stationary phase	Mobile phase	Ref.
DADMAC	Quaternized Styragel	acid medium	[146]
DADMAC	G 3000 PW G 5000 PW	0.1 M NaCl 0.2 M NaCl	[147, 148]
DADMAC	Superose	0.4 M NaCl/Na-ac (9:1) pH=5.5	[142]
DADMAC	Shodex OH pak B-806	0.5 M NaCl	[149]
DADMAC	Fractogel TSK-55(S)	0.5 M NaCl, 0.5 M $NaNO_3$	[74]
DADMAC	Superose-6	0.5 M NaCl pH= 6.5, 0.25 M NaOAc buffer	[136]
DADMAC/ AAM	Combination of 5 columns: poly(glycidyl methacrylate) gel: 1. guard, 2. 6000, 3. 5000, 4. 3000 and hydroxyethyl methacrylate gel: 5. 40	0.5 M $NaNO_3$ with 10 ppm NaN_3	[67]
DADMAC/ NMVA	Progel-TSK-PW	0.5 M $NaNO_3$	[45]

samples varied. This can be expected for technical products with broad molar mass distribution. The highest molar mass of PDADMAC measured by SEC was $M_w \approx 8 \times 10^5$ g mol^{-1}, $M_w / M_n \approx 2.4$ [137].

6.3.2
Other Methods

Molar mass distributions of PDADMAC were also determined by fractionation using dioxane/methanol system [144] and from sedimentation velocity measurements in 1 m NaCl solution [134]. The molar mass calculations were based on the s-M relation [134]:

$$s_0 [s] = 6.1 \cdot 10^{-16} M_w^{0.47} \qquad (25)$$

Information on polydispersity of PDADMAC samples synthesized under different conditions could be obtained further by the M_z/M_w-ratio from low-speed sedimentation equilibrium experiments in 0.5 m NaCl [35].

Recently, Kulicke et al. [145] applied the principle of Flow-Field-Flow-Fractionation (F^4) in combination with multi angle laser light scattering (MALLS) to characterize PDADMAC. Differential and integral molar mass distributions have been published for PDADMAC. One advantage of this method seems to be that it can be extended to the investigation of the mass distribution of aggregated and particulate components.

6.4
Determination of Structural Nonuniformities

As demonstrated in Figs. 6 and 7, and discussed in Sect. 3, PDADMAC products can contain structured chains. By specific addition of MTAAC [44] or by initiation with multifunctional starting agents [35] the application properties can be improved [44]. Therefore, not only the molar masses and the molar mass distributions are of interest but also information about the portion of branched structures. Based on differences of the partial specific volumes of linear and branched PDADMAC a preparative separation was possible by preparative ultracentrifugation using a density gradient method [35]. Figure 25 shows the composition analysis for one polymer sample containing linear and branched moieties [35].

Although this method is very time consuming, it can, in principle, be regarded as an alternative to the chromatographic cross fractionation to characterize copolymers and other complicated polyelectrolyte structures.

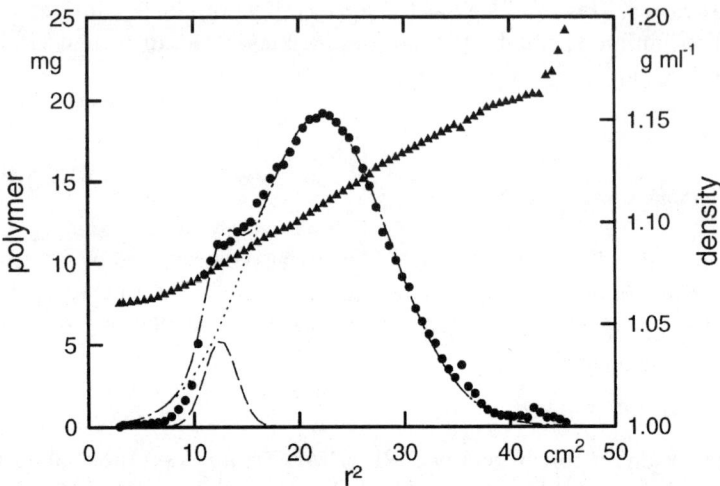

Fig. 25. Composition analysis of technical PDADMAC (● experimental values; –·–·– Gaussian fit; ··· linear fraction; – – – branched fraction; ▲ density) (Data taken from [35])

7
Interactions of Diallyldimethylammonium Chloride Polymers in Solution and at Interfaces

Interactions of polyelectrolytes in solution and at interfaces are of predominantly Coulombic nature [27, 150]. However, other interactions including dipol-dipol, charge-transfer, and hydrophobic effects also have to be considered [27, 151, 152]. Table 13 demonstrates general characteristics of interactions of polycations with oppositely charged species which can be applied to PDADMAC.

These solution based interactions are relevant to most technical applications of PDADMAC and copolymers (see Sect. 8).

7.1
Interactions with Low Molecular Mass Components

Anions

The theoretical bases for interaction with counterions have been discussed in Sect. 5.1. However, these theories do not take into account the special nature of the ions and only the valency was considered [102–108, 114–117]. The solubility caused by the variation in the counterions cannot, for example, be predicted. According to [35] the solubility of poly(diallyldimethylammonium halides) decreases in the order $Cl^- > Br^- > I^-$. The homopolymer precipitates in the iodide form. This is in contrast to ammonium halides where the iodide has the highest solubility [153]. Comparing the effects of Cl^-, SO_4^{2-}, and PO_4^{3-} a very similar in-

Table 13. Interactions of PDADMAC in Solution and at Interfaces

Interaction with	Effect
Anions	counterion binding, counterion exchange, aggregation, precipitation
Anionic surfactants	complexation, charge neutralization, aggregation, precipitation, structure formation
Polyanions	complexation, charge neutralization • colloidal aggregates • stable turbid dispersions • flocculated precipitates • coherent gels
Anionic surfaces	• flat adsorption (DADMAC with low or moderate molar mass) • partial adsorption (copolymers with low charge density, homo- and copolymers with high molar mass, branched structures) • recharge of surfaces, neutralization, stabilization

fluence has been detected by viscosity measurements indicating that there is virtually no change of the solution state [154]. However, a strong decrease of the intrinsic viscosity and precipitation was observed in aqueous PDADMAC solutions in the presence of $[Fe(CN)_6]^{4-}$ and $[Fe(CN)_6]^{3-}$ [154].

Surfactants

Several models of complex formation between polyelectrolytes and ionic surfactants have been proposed [155–159]. However, no generally accepted model is currently available [27]. With PDADMAC, both water soluble [156–158] and water insoluble complexes [159] have been reported. Generally, highly ordered structures are formed. Several groups have examined the use of polycations with surfactants to stabilize surfactant films to act as model cell membranes [159]. By combining lecithin with PDADMAC, model membranes were produced which mimic several cell membrane physical properties [160, 161]. Highly ordered materials with ultra-low surface energies could be created by complexation of PDADMAC with fluorinated surfactants. PDADMAC was chosen since it combines good accessibility with an appropriate charge distance along the polymer backbone [162].

7.2
Interactions with Oppositely Charged Polymers

The polyanion-polycation complex (symplex) formation process is a phenomenon that had long been known on an empirical base from the mutual precipitation of proteins [150]. The internal structure and the properties of the resulting complexes are strongly influenced by the nature of the polymeric components and the system conditions. The polymer parameters include the molar mass, the charge density/charge density distribution, the nature of the ionic group, or the chain architecture. Likewise, the concentration range, the pH, and the ionic strength are important [27, 150, 163-165)]. As given in Table 13, different degrees of aggregation can be produced including quasi-soluble particles on a colloidal level and phase separation. Furthermore, the phase separation can occur as hard precipitation or membrane formation. All borderline cases have been observed for PDADMAC [27, 150, 166-169].

While for the complexation with poly(sodium styrene sulfonate) or sodium cellulosesulfate 1:1 stoichiometry has been reported [150] a non-stoichiometric complex results with sodium carboxymethylcellulose [150]. Optimized conditions make it possible to create membranes with various properties using the PDADMAC/sodium cellulosesulfate system [166-168]. However, the symplex formation with PDADMAC or copolymers mostly results in flocculated precipitates [27, 150, 169]. Highly ordered mulilayer assemblies were prepared by alternate reaction of PDADMAC and various polyanions [170, 171]. Recently, the efficiency and selectictivity of protein separation via PEL coacervation were examined using PDADMAC [172].

7.3
Interactions at Surfaces

Charged and uncharged surfaces may exist as surfaces of particles having different size and as solid surfaces. Though there are different dimensions ranging from colloidal particles to macroscopic areas, the same principal considerations are valid [27]. Various models have been developed to describe the interactions between surfaces and polyelectrolytes [173, 174]. Interactions between DADMAC polymers and negatively charged particles or surfaces are important in regard to many practical applications. These applications include separation, stabilization, solubilization, flocculation processes, and surface modification. Application examples for these most important interactions will be discussed in the following section.

8
Applications of Diallyldimethylammonium Chloride Homo- and Copolymers

PDADMAC was initially produced in order to provide specialty papers with high electroconductivity [23]. It has since 1966 been increasingly applied in many industrial fields. The IBM Patent Server reported more than 120 patents in U.S. related to the application of PDADMAC or copolymers. Over 30 of these have been registered since 1995, indicating the emerging applications for PDADMAC. A listing of applications has been presented in [1, 15, 18] and emerging applications including advanced materials with novel properties are reported in [175]. The present review does not intend to give listing of all known applications. However, Table 14 summarizes application fields and general applications for homo- and copolymers of DADMAC. Detailed explanations of established and most important applications follow in the next Sections.

Table 14. Applications of DADMAC homo- and copolymers

Application field	Application
Paper and textile production	retention, coagulation, flocculation, wet strength, dry strength improvement, dewatering, color fastness, dye fixation, antistatic agents, antimicrobial treatment
Water / waste water treatment	primary coagulation, flocculation, sludge dewatering, demulsification
Coal, mineral, glass industries	flotation, flocculation, stabilization, dewatering, hydrophilization, viscosity reduction
Cosmetics, hair treatment	components in: hair conditioners, hair rinses, shampoos, deodorants, liquid soaps, antistatic, antimicrobial components
Biological, medical, food processes	virus removal, insecticides, algaecides, conservation, immobilization, cell fixation
Miscellaneous	membrane modification, film modification, antifouling compositions, analytical agents

8.1
Paper Manufacturing

8.1.1
Retention and Drainage Agents

Paper machines are devices for continuously forming, dewatering, pressing and drying a web of paper fibers [176]. Basically, a dilute suspension of fibers is applied to a continuous wire screen which can have several configurations. Retention is a measure of how much material remains on the paper machine wire and is incorporated into the final sheet. Retention occurs by filtration (fibers and fines larger than 200 µm or the size of the largest openings of the paper machine) and adsorption onto the fibers via the formation of secondary chemical bonds (particles less than 10 µm). Drainage implies the removal of water from the paper machine. Higher drainage rates mean faster paper machine operation and/or energy savings in the dryer section. Copolymers of acrylamide and DADMAC are widely used for both retention and drainage agents [176, 187]. Other uses of DADMAC copolymers with acrylamide as retention and drainage agents are mentioned in the patent literature [179,188]. A moderately branched copolymer of acrylamide and DADMAC is claimed to have improved retention in the manufacture of paper and cardboard [189].

8.1.2
Wet Strength Additives

Wet strength is a desirable attribute in many paper products including napkins, paper towels, household tissues and disposable hospital wear all of which come in contact with water during their use [176, 186]. Typically, an untreated cellulose fiber would lose 95–97% of its strength when saturated with water. To improve the wet strength, paper producers have, for years, used chemicals during the paper manufacturing process. Wet strength additives function by forming covalent bonds between fibers and by generating their own crosslinked network. The dry strength of the paper product will increase as well. Typically these materials are defined either as permanent or "temporary" wet strength agents. Permanent wet strength agents will provide a product which retains more than 50% of its original wet strength after a 5 minutes exposure to water. On the other hand, temporary wet strength resins will produce paper products retaining less than 50% of its original wet strength after the same exposure period [176]. Some temporary wet strength agents are based on glyoxylated acrylamide-DADMAC copolymers as in the case of some commercial products (Parez 631 NC, manufactured by Cytec) and described by Coscia [177] and Williams [178]. The use of DADMAC in the preparation of temporary wet strength agents is further mentioned in the patent literature [179–186].

8.1.3
Other Paper Manufacturing Agents

One common problem in the paper manufacture is the deposition of wood pitch on the pulp handling equipment or in the paper machine [176]. The pitch is liberated from the fibers during the fabrication of paper and tends to accumulate as a colloidal suspension of negatively charged particles. Later, these particles cause problems by filling the wire of the paper machine, producing holes in the finished product or by collecting on the felts or machine parts as sticky lumps. These problems can be controlled by the use of copolymers of DADMAC and acrylamide [189] and DADMAC and vinyl alkoxysilanes [86, 191]. Other additives used for this application include a hydrophobic polyelectrolyte copolymer based on DADMAC [192]. A synergistic approach for pitch control is reported by using water-soluble zirconium carbonate in combination with PDADMAC [193].

Homopolymers of DADMAC and copolymers of acrylamide and DADMAC are applied in the coagulation of fibers recycled from coated broke, a term used to describe a paper which cannot be sold for different reasons [176]. It is claimed that polymer solutions and water-in-oil emulsions containing DADMAC offer both superior performance and cost effectiveness compared to the traditional polymer additives used for this purpose [190, 194].

8.2
Mining Industry

Homo- and copolymers of DADMAC are widely used for solid/liquid separations (dewatering) of different slurries in the mining industry. Specific examples of the minerals treated with these polymers include coal, taconite, trona, sand, gravel and titania slurries. In coal production, several applications of DADMAC and DADMAC copolymers are reported in the patent literature. The claims in these patents include: i) the recovery of clean coal and the reduction of ash content through the use of PDADMAC [195], ii) the separation of gangue from coal in a coal refuse slurry in a multi-stage separation process using PDADMAC [196] and iii) a method for dewatering coal tailings and clean coal products by the use of copolymers of DADMAC and vinyl trialkoxysilane [197]. Pilot studies in the brown coal industry performed with branched PDADMAC indicated an improved dewatering performance over linear PDADMAC and other anionic and cationic acrylic polymers [218]. Improved dewatering of waste solids generated in common mining processing operations is claimed by the use of hydrophobically modified copolymer of DADMAC and acrylic monomers [198], DADMAC-vinyl trialkoxysilane copolymers [199, 200] and copolymers of DADMAC and acrylamide crosslinked with triallyl amine [201]. Dust suppressants for finely divided mineral particles are obtained with dilute DADMAC homopolymer solutions [202]. Copolymers of acrylamide and DADMAC obtained by inverse-emulsion copolymerization are reported to improve dewatering aid in mineral processing [203].

8.3
Water Treatment Industry

It was previously mentioned that PDADMAC (Cat-Floc) was the first commercial flocculant approved for potable water [26]. Since then, PDADMAC has been continuously used for coagulation/flocculation both in potable water and waste water treatment. A good example of the performance of PDADMAC in the coagulation of colloidal solids is the reduction of turbidity in fresh water of 150 mg L^{-1} of $Ca(OH)_2$. A reduction of 82% in turbidity is observed with the addition of only 2 mg L^{-1} of branched PDADMAC [217]. In addition, PDADMAC and copolymers of DADMAC are reported to be effective in the removal of hard-to-eliminate impurities in the water treatment industry. Emulsified impurities from streams of a petroleum refinery waste water and an automotive oily effluent water have been removed by the use of water soluble copolymers consisting essentially of DADMAC and small amounts of anionic acrylic monomers [89].

Copolymers of DADMAC with acrylamide in combination with several coalescent agents have been reported to improve the water quality of overboard waters discharged from off-shore oil producing rigs [204]. A method for removing oil and metal ions is reported by the addition of a polymer formed by condensing tannin with PDADMAC [205]. A complex system composed of inorganic salts and organic adduct polymers based on DADMAC and other monomers has been reported. These systems have been found to be effective for the detackification and clarification of acid and alkaline paint and lacquer waste waters as well as spray booth wastes [206]. Hydrophobically associating copolymer compositions based on water soluble silicon and DADMAC copolymers are reported in the efficient removal of fat, blood, tissue and other solids from food and biological processing wastes [84]. Copolymers of vinyltrimethoxysilane and DADMAC have successfully been applied for the clarification of ink-containing waste waters from recycled paper production [207]. A system based on a hydrophobically-modified copolymer of DADMAC with acrylic monomers in combination with other acrylic flocculants is used for the dewatering of industrial sludges [208]. Other water treatment applications of DADMAC polymers include the stabilization of metal ions with terpolymers containing styrene sulfonic acid and DADMAC [88], the inhibition of scale by the use of terpolymers containing vinyl alcohol and DADMAC [83], the preparation of oil-in-water breakers with an hydrophobically-modified copolymer based on DADMAC and other acrylic monomers [209], the demulsification of oily waste water using copolymers of vinyl alkoxysilanes and DADMAC [210] and a method of cleaning waste waters from laundries with the use of PDADMAC [85].

8.4
Miscellaneous Applications

A novel capillary electrophoresis method using solutions of non-crosslinked PDADMAC is reported to be effective in the separation of biomolecules [211]. Soil studies conducted with PDADMAC report the minimization of run-off and erosion of selected types of soils [212]. In similar studies, PDADMAC has found to be a good soil conditioner [213]. The use of PDADMAC for the simultaneous determination of inorganic ions and chelates in the kinetic differentiation-mode capillary electrophoresis is reported by Krokhin [214]. Protein multilayer assemblies have been reported with the alternate adsorption of oppositely charged polyions including PDADMAC. Temperature-sensitive flocculants have been prepared based on n-isopropylacrylamide and DADMAC copolymers [215]. A potentiometric titration method for the determination of anionic polyelectrolytes has been developed with the use of PDADMAC, a marker ion and a plastic membrane. The end-point is detected as a sharp potential change due to the rapid decrease in the concentration of the marker due to its association with PDADMAC [216].

9
Conclusions

This review demonstrated that research on diallyldimethylammoium chloride and its polymers have contributed to the general understanding of the polymerization of ionic monomers, the development of methods for the molecular characterization possibilities of cationic polyelectrolytes, and the understanding regarding polyelectrolyte behavior. However, in comparison to the industrial importance of diallyldimethylammonium chloride polymers, the level of fundamental knowledge is far from adequate. In particular, copolymerization processes with monomers other than acrylamide, the characterization of copolymers related to their chain architecture and charge distribution, the dependence of polyelectrolyte interactions on the chemical structure and the medium, and the structure formation in concentrated systems remain to be elucidated. Since polyelectrolytes are currently emerging as an important and challenging topic in polymer science, it is expected that the interest in polymers based on diallyldimethyldiammonium chloride will not diminish over the coming two decades. Moreover, the recently proliferating applications will likely lead to the development of additional novel DADMAC based materials and associated studies into their characterization and solution behavior.

10
References

1. Butler GB (1992) Cyclopolymerization and cyclocopolymerization, Marcel Dekker, New York
2. Butler GB, Zhang NZ (1991) In: Water soluble polymers, ACS Symposium Series 467, Washington DC, 25
3. Butler GB, Angelo RJ (1957) J Am Chem Soc 79:3128
4. Butler GB (1966) U.S. Pat 3,288,770
5. Butler GB, Ingley FL (1951) J Am Chem Soc 72:894
6. Wandrey C, Jaeger W, Reinisch G (1981) Acta Polym 32:197
7. Wandrey C, Jaeger W, Reinisch G (1981) Acta Polym 32:257
8. Topchiev DA, Nashmetdinova GT (1983) Vysokomol Soed A25:636
9. Topchiev DA, Nashmetdinova GT, Kartashewskij AI, Netshaeva AV, Kabanov VA, (1983) Isv Akad Nauk SSR, Ser Khim: 2232
10. Hahn M, Jaeger W, Wandrey C, Reinisch G (1984) Acta Polym 35:350
11. Jaeger W, Hahn M, Wandrey C, Seehaus F, Reinisch G (1984) J Macromol Sci Chem A21:593
12. Kabanov VA, Topciev DA (1988) Vysokomol Soed A30:675
13. Huang PC, Reichert KH (1988) Angew Makromol Chem 162:19
14. Huang PC, Reichert KH (1989) Angew Makromol Chem 165:1
15. Jaeger W, Gohlke U, Hahn M, Wandrey C, Dietrich K (1989) Acta Polym 40:161
16. Jaeger W, Hahn M, Wandrey C (1989) In: Polymer reaction engineering. Reichert KH, Geiseler W (eds) VCH Weinheim:239
17. Lancaster JE, Bacchei L, Panzer HP (1976) J Polym Sci, Polym Lett 14:549
18. Ottenbrite RM, Shillady DD (1980) In: Polymeric amines and ammonium salts. Goethals E (ed) Pergamon Press, Oxford:143
19. Wandrey C, Jarger W, Reinisch G, Hahn M, Engelhard G, Jancke H, Ballschuh D (1981) Acta Polym 32:179
20. Hoover MF (1970) J Macromol Sci-Chem A4:1327
21. Ottenbrite RM, Ryan WS (1980), Ind Eng Chem Prod Res Dev 19:529
22. Butler GB (1980) In: Polymeric amines and ammonium salts. Goethals E (ed) Pergamon Press, Oxford:125
23. Hoover FM, Carr HE (1968) Tappi 51:552
24. Nicke R (1982) Zellstoff Papier 31:19
25. Jaeger W, Hong LT, Philipp B, Reinisch G, Wandrey C (1979) Zellstoff Papier 28:268
26. Wandrey C, Jaeger W, Starke W, Wotzka J (1984) Wasserwirtschaft Wassertechnik 34:18
27. Dautzenberg H, Jaeger W, Kötz J, Philipp B, Seidel C, Stscherbina D (1994) Polyelectrolytes. Formation, characterization, application. Carl Hanser Verlag, München
28. Mandel M (1988) In: Encycl Polym Sci Eng vol 11: Wiley & Sons, New York: 739
29. Anonymous (1968) Chem Eng News: 46

30. Negi Y, Harada S, Hishizuka O (1967) J Polym Sci A1, 5:1951
31. Harada S, Arai K (1967) Makromol Chem 107:64
32. Wandrey C (1980) Dissertation, FB Chemie AdW, Berlin
33. Ballschuh D, Jaeger W, Hahn M, Reinisch G, Ohme R (1977) DD-PS 128 392
34. Hunter WE, Sieder TP (1979) U.S. Pat 4,151,202
35. Wandrey C, Görnitz E (1992) Acta Polym 43:320
36. Ohme R, Ballschuh D, Seibt H (1991) Tenside Surf. Det. 28:180
37. Martin V, Ringsdorf H, Thunig D (1977) In: Polymerization of organized systems. Midland Macromolecular Monographs, Gardon & Breack Sci Publ 3:175
38. Wandrey C (1997) Polyelektrolyte-Makromolekulare Parameter und Elektrolytverhalten. Cuvillier Verlag, Göttingen
39. Wedler G (1987) Lehrbuch der Physikalischen Chemie. VCH, Weinheim
40. Butler GB (1960) J Polym Sci 48:279
41. Masterman TC, Dando NR, Weaver DG, Seyferth D (1994) J Polym Sci, Part B: Polym Physics 32:2263
42. Wehrman C (1974) Diploma Thesis, TU Dresden
43. Birnstein D (1975) Diploma Thesis, TU Dresden
44. Jaeger W, Wandrey C, Hahn M, Ballschuh D, Ohme R, Staeck R, Biering H (1987) EP 0,264,710
45. Ruppelt D, Kötz J, Jaeger W, Friberg SE, Mackay RA (1997) Langmuir 13:3316
46. Brand F (1995) Thesis, TU Berlin
47. Brand F, Dautzenberg H, Jaeger W, Hahn M (1997) Angew Makromol Chem 248:41
48. Truong ND, Galin JC, Francois J (1986) Polymer 27:459
49. Vu C, Cabestany J (1991) J Appl Polym Sci 42:2857
50. Murphy PD, Pietro RAD, Lund CJ Weber WD (1994) Macromolecules 24:279
51. Hahn M, Jaeger W (1992) Angew Makromol Chem 198:165
52. Jaeger W, Hahn M, Lieske A, Zimmermann A (1996) Macromol Symp 111:95
53. Hahn M (1983) Dissertation, FB Chemie AdW, Berlin
54. Martynenko AI, Wandrey C, Jaeger W, Hahn M, Topchiev DA, Reinisch G, Kabanov VA (1985) Acta Polym 36:516
55. Hunkeler D (1991) Macromolecules 24:2160
56. Jaeger W, Hong LT, Philipp B, Reinisch G, Wandrey C (1980) In: Polymeric amines and ammonium salts. Goethals E (ed) Pergamon Press, Oxford: 155
57. Kabanov VA, Topchiev DA, Nazhmetdinova GT (1983) Izv Akad Nauk SSSR Ser Khim 9:2146
58. Babaev HA, Martynenko AI, Topchiev DA, Kabanov VA, Wandrey C, Hahn M, Jaeger W, Reinisch G (1985) Acta Polym 36:396
59. Wyroba A (1981) Polymeri 26:139
60. Hahn M, Jaeger W, Reinisch G (1983) Acta Polymer 34:322
61. Schuller WHJ, Price JA, Moore ST Thomas WM (1959) J Chem Eng Data 4:273
62. Mayo FR, Lewis FM (1944) J Am Chem Soc 66:1595
63. Riggs JP, Podriguez FJ (1967) J Polym Sci A1 5:3151
64. Wandrey C, Jaeger W (1985) Acta Polym 36:100
65. Tanaka H (1986) J Polym Sci Polym Chem Ed 24:29
66. Matsumoto A, Wakabayashi S, Oiwa M, Butler GB (1989) J Macromol Sci Chem A26:1475
67. Brand F, Dautzenberg H, Jaeger W, Hahn M (1997) Angew Makromol Chem 248:41
68. Huang PC, Singh P, Reichert KH (1986) In: Polymer Reaction Engineering. Reichert KH, Geiseler W (eds) VCH Weinheim: 125
69. Singh P (1986) DFG Report 223/13-1. Inst Techn Chem TU Berlin
70. Hunkeler D, Hamielec AE (1991) Polymer 32:2626
71. Kelen T, Tüdös F (1975) J Macromol Sci Chem A9:1
72. Reilly PM, Blau GE (1974) Can J Chem Eng 52:289
73. Baade W, Hunkeler D, Hamielec AE (1988) J Appl Polym Sci 38:185

74. Wandrey C, Görnitz E (1995) Polym News 22:377
75. Janietz S, Hahn M, Jaeger W (1992) Acta Polym 43:230
76. Schuller WH, Thomas WM, Moore ST House RA (1959) U.S. Pat 2,884,054
77. Nixon AC, Berrigan PJ, Williams PH (1966) US Pat: 3,278,474
78. Deng YL, Pelton R (1995) Macromolecules 28:4617
79. Moench D, Hartmann H, Buechner KH (1990) OS DE 3,909,005
80. Jen Y (1960) U.S. Pat 2,958,673
81. Harada S, Katayama (1968) U.S. Pat 3,375,233
82. Bhattacharyya BR, Dalsin PD (1987) U.S, Pat 4,713,431
83. Amjad Z, Masler, WF (1989) U.S. Pat 4,889,637
84. Chung DK, Ramesh M (1997) U.S. Pat 5,597,490
85. Tibbits D, (1996) U.S. Pat 5,529,696
86. Shetty, CS, Ramesh M (1996) U.S. Pat 5,527,431
87. John St MR, Alexander RL (1990) U.S. Pat 5,013,456
88. Amjad Z, Masler WF (1989) U.S. Pat 4,885,097
89. Bhattacharyya BR, Srivatsa SR, Dwyer ML (1987) U.S. Pat 4,715,962
90. Gopoakrishnan S, Butler BG, Hogen-Esch TE, Zhang NZ (1991) ACS Symp Ser 467:175
91. Butler GB, Do CH (1991) ACS Symp Ser 467:151
92. Kevelam J, Engberts BFN (1995) Langmuir 11:793
93. Topchiev DA, Sivov NA, Goethals EJ (1994) Russ Chem Bull 43:1864
94. Becker LW, Larson EH (1986) U.S. Pat 4,617,362
95. Polyelectrolytes: Science and technology. Harada M ed (1993) Marcel Decker, New York
96. Schmitz KS (1993) Macroions in solution and colloidal suspension, VCH, New York
97. Förster S, Schmidt M (1995) Adv Polym Sci 120:51
98. Fuoss RM, Katchalsky A, Lifson S (1951) Proc Acad Sci 37:579
99. Alexandrowicz Z, Katchalsky A (1963) J Polym Sci A1:3231
100. Oosawa F (1971) Polyelectrolytes, Marcel Dekker, New York
101. Manning GS Zimm BH (1965) J Chem Phys 43:4250
102. Manning GS (1965) J Chem Phys 43:4260
103. Manning GS (1969) J Chem Phys 51:924
104. Manning GS (1969) J Chem Phys 51:934
105. Manning GS (1977) Biophys Chem 7:95
106. Manning GS (1978) Biophys Chem 9:65
107. Manning GS (1978) Quart Rev Biophys 11:179
108. Manning GS (1979) Acc Chem Res 12:443
109. Gennes de PG (1979) Scaling conceps in polymer physics, Cornell Univ Press, Ithaca, New York
110. Odijk T (1977) J Polym Sci Polym Phys Ed 15:477
111. Odijk T (1979) Macromolecules 12:688
112. Odijk T, Houwaart AC (1978) J Polym Sci, Polym Phys Ed 16:627
113. Skolnik J, Fixman M (1977) Macromolecules 10:944
114. Iwasa K, Quarrie Mc DA, Kwak JCT (1978) J Phys Chem 82:1979
115. Gueron M, Weisbuch G (1979) J Phys Chem 83:1991
116. Yoshida N (1978) J Chem Phys 69:4867
117. Yoshida N (1982) Chem Phys Lett 90:207
118. Leeuven v HP, Cleven FRM, Valenta P (1991) Pure Appl Chem 63:1251
119. Manning GS (1970) Biopolymers 9:1543
120. Huzienga JR, Grieger PF, Wall FT (1950) J Am Chem Soc 72:2636
121. Eisenberg H (1958) J Polym Sci 30:47
122. Kuruczev T, Steel BJ, (1967) Rev Pure Appl Chem 17:149
123. Manning GS (1975) 79:262
124. Schmitz KS ed (1994) Macro-ion characterization. ACS Symp Ser 548, Washington
125. Oehme F (1986) Ionenselektive Elektroden. Hüthig, Heidelberg

126. Rios HE, Gamboa C, Ternero G (1991) J Polym Sci, Part B: Polym Phys 29:805
127. Kwak JCT (1973) J Polym Sci Chem 77:2790
128. Vink H (1982) Makromol Chem 183:2273
129. Wandrey C, Zarras P, Vogl O (1995) Acta Polym 46:247
130. Wandrey C (1996) Ber Bunsenges Phys Chem 100:869
131. Görnitz E, Hahn M, Jaeger W, Dautzenberg H (1997) Progr Colloid Polym Sci 107:127
132. Burkhardt CW, McCarthy KJ, Parazak DP (1987) J Polym Sci, Polym Lett Ed 25:209
133. Wandrey C, Jaeger W, Reinisch G (1982) Acta Polym 33:156
134. Timofejeva GI, Pavlova SA, Wandrey C, Jaeger W, Linow KJ, Görnitz E (1990) Acta Polym 41:479
135. Dubin PL, The SS, Gan LM, Chew CH (1990) Macromolecules 23:2500
136. Xia J, Dubin PL, Edwards S, Havel H (1995) J Polym Sci, Part B: Polym Phys 33:1117
137. Dautzenberg H, Görnitz E, Jaeger W (1998) Macromol Chem Phys 199:1561
138. Dautzenberg H, Rother G (1991) J Appl Polym Sci, Appl Polym Symp 48:351
139. Bekturov EA, Kudaibergenov SE, Ushanov VZ, Saltybaeva SS (1985) Makromol Chem 186:71
140. Okay O, Schindlbauer H (1984) Angew Makromol Chem 122:21
141. Janado M (1988) In: Aqueous size exclusion chromatography. Dubin PL (ed), Elsevier, Amsterdam
142. Strege HA, Dubin PL (1989) J Chromatogr 463:165
143. Bruessau RJ (1992) Makromol Chem Makromol Symp 61:199
144. Wandrey C, Jaeger W, Reinisch G (1982) Acta Polym 33:442
145. Thielking H, Adolphi U, Kulicke WM (1996) Nachr Chem Tech Lab 44:370
146. Butler GB, Wu C (1976), U.S. Pat 3,962,206
147. Dubin PL, Levy IJ (1982) J Chromatogr 235:377
148. Levy IJ, Dubin PL (1982) Ind Eng Chem Prod Res Dev 21:59
149. Kokufuta E, Takahashi K (1986) Macromolecules 19:351
150. Philipp B, Dautzenberg H, Linow KJ, Kötz J, Dawydoff W (1989) Progr Polym Sci 14:91
151. Philipp B, Kötz J, Linow KJ, Dautzenberg H (1991) Polym News 16:106
152. Bowman WA, Rubinstein M, Tan JS (1997) Macromolecules 30:3262
153. Tabellenbuch Chemie (1968) Dt Verlag Grundstoffind, 288
154. Bekturov EA, Kudaibergenov SE, Ushanov VZ, Saltybaeva SS (1985) Makromol Chem 186:71
155. Wei YC, Hudson SM (1995) J Macromol Sci, Rev Macromol Chem Phys C35:15
156. QuiggMc DW, Kaplan JI, Dubin PL (1992) J Phys Chem 96:1973
157. Xia J, Zhang H, Rigsbee DR, Dubin PL, Shaikh T (1993) Macromolecules 26:2759
158. Swanson-Vethamuthu M, Dubin PL, Almgren M, Li, Y (1997) J Colloid Interface Sci 186:414
159. Ober CK, Wegner G (1997) Adv Mater 9:17
160. Antonietti M, Kaul A, Thünemann A (1995) Langmuir 11:2633
161. Antonietti M, Wenzel A, Thünemann A (1996) Langmuir 12:2111
162. Antonietti M, Henke S, Thünemann A (1996) Adv. Mater 8:41
163. Mikulik J, Vinklarek Z, Vondruska M (1993) Collect Czech Chem Commun 58:713
164. Dautzenberg H, Koetz J, Linow KJ, Philipp B, Rother G (1994) In: Macromolecular complexes in chemistry and biology. Dubin PL (ed) Springer, Berlin, New York
165. Kabanov VA (1994) In: Macromolecular complexes in chemistry and biology. Dubin PL (ed) Springer, Berlin, New York
166. Dautzenberg H, Loth F, Fechner K, Mehlis B, Pommerenning K (1985) Macromol Chem Suppl 9:203
167. Dautzenberg H, Lukanoff B, Eckert U, Tiersch B, Schuldt U (1996) Ber Bunsenges Phys Chem 100:1045
168. Schwarz HH, Richau K, Paul D (1992) Polym Bull 25:95
169. Dautzenberg H, Hartmann, Grunewald S, Brand F (1996) Ber Bunsenges Phys Chem 100:1024

170. Decher G, Schmitt J (1992) Progr Colloid Polym Sci 89:160
171. Lvov Y, Ariga K, Ichinise I, Kunitake T (1996) Thin Solid Films 284/285:797
172. Wang Y, Gao, JY, Dubin PL (1996) Biotechnol Prog 12:356
173. Fleer GJ, Cohen Stuart MA, Scheutchens JMHM, Cosgrove T, Vincent B (1993) Polymers at Interfaces, Capman & Hall, London
174. Horn D, Linhart F (1991) In: Paper chemistry. Roberts CJ (ed) Blackie, Glasgow
175. Wandrey C (1998) Polym News 23:123
176. Biermann CJ (1996) Handbook of pulping and papermaking, Academic Press, San Diego
177. Coscia AT, Williams LL (1971) U.S. Pat 3,556,932
178. Williams LL, Coscia AT (1971) U.S. Pat 3,556,933
179. Ballweber EG, Jansma RH, Phillips KG (1980) U.S. Pat 4,217,425
180. Bjorkquist DW, Schmidt WW (1986) U.S. Pat 4,603,176
181. Furman, Jr. GS (1993) U.S. Pat 5,187,219
182. Noda, I (1993) U.S. Pat 5,200,036
183. Dauplaise DL, Kozakiewicz JJ, Schmitt JM (1994) U.S. Pat 5,320,711
184. Darlington WB, Lanier WG (1995) U.S. Pat 5,427,652
185. Darlington WB, Lanier WG (1995) U.S. Pat 5,466,337
186. Jansma RH, Begala AJ, Furman GS (1996) U.S. Pat 5,490,904
187. Unbehend JE, Britt KW (1981) In: Pulp and paper. Chemistry and chemical technology, vol.3. James P. Casey (ed), John Wiley&Sons, New York
188. Lim SK, Bloomquist AE, Shaper RJ (1978) U.S. Pat 4,077,930
189. Hund R, Philibert E (1995) U.S. Pat 5,393,381
190. St. John MR (1992) U.S. Pat 5,131,882
191. Shetty CS, Ramesh M (1996) U.S. Pat 5,510,439
192. Finck MR, Greer CS, Ramesh, M (1993) U.S. Pat 5,246,547
193. Greer CS, James NP (1993) U.S. Pat 5,230,774
194. Hurlock JR, Ballweber EG, Connelly LJ (1975) U.S. Pat 3,920,599
195. Antonetti JM, Snow GF (1979) U.S. Pat 4,141,691
196. Sykes RC, Connelly LJ, Roe WJ (1985) U.S. Pat 4,555,329
197. Kerr EM, Ramesh M (1997) U.S. Pat 5,622,647
198. Kerr EM (1995) U.S. Pat 5,603,841
199. Kerr EM, Ramesh M (1995) U.S. Pat 5,476,522
200. Kerr EM, Ramesh M (1997) U.S. Pat 5,597,475
201. Pillai KJ, Kerr EM (1996) U.S. Pat 5,518,634
202. Tippett JM, Groesz M (1993) U.S. Pat 5,215,784
203. Richardson PF, Bhattacharyya, BR (1987) U.S. Pat 4,673,511
204. Marble RA, Braden ML (1992) U.S. Pat 5,128,046
205. Meyer EM, Wood, MR (1993) U.S. Pat 5,256,304
206. Waldmann JJ (1994) U.S. Pat 5,294,352
207. Chung DK, Ramesh M (1997) U.S. Pat 5,624,569
208. Chung DK, Shetty CS, Ramesh M (1997) U.S. Pat 5,601,725
209. Ramesh M, Sivakumar A (1997) U.S. Pat 5,635,112
210. Sivakumar A, Ramesh M (1996) U.S. Pat 5,560,832
211. Demorest DM, Werner WE, Wiktorowics JE, (1993) U.S. Pat 5,264,101
212. Bernas SM, Oades JM, Churchman GJ (1995) Australian J Soil Research 33:805
213. Bernas SM, Oades JM, Churchman GJ, Grant CD (1995) Australian J Soil Research 33:369
214. Krokhin OV, Hoshino H, Shpigun OA, Yotsuyanagi T (1997) J Chromatogr 776:329
215. Deng YL, Xiao HN, Pelton R (1996) J Colloid Interface Sci 179:188
216. Masadome T, Imato T (1995) Fresenius J Analyt Chem 352:596
217. Philipp B, Jaeger W, Gohlke U, Wandrey C, Hahn M, Dietrich K, Reinisch, G, Linow, KJ, Dautzenberg H, Kötz J (1986) Paperi ja Puu 5:419
218. Philipp B, Gohlke U, Jaeger, H, Kötz J (1989) Wiss Fortschritt 39:137

219. Mortimer DA (1991), Polym Internat 25:29
220. Itaya T, Kawabata Y, Ochiai H, Ueda K, Imamura A (1994) Bull Chem Soc Jpn 67:2047

Editor: Prof. H.-H. Kausch and Prof. K.-S. Lee
Received: May 1998

Author Index Volumes 101–145

Author Index Volumes 1–100 see Volume 100

de, Abajo, J. and *de la Campa, J.G.*: Processable Aromatic Polyimides. Vol. 140, pp. 23-60.
Adolf, D. B. see Ediger, M. D.: Vol. 116, pp. 73-110.
Aharoni, S. M. and *Edwards, S. F.*: Rigid Polymer Networks. Vol. 118, pp. 1-231.
Améduri, B., Boutevin, B. and *Gramain, P.*: Synthesis of Block Copolymers by Radical Polymerization and Telomerization. Vol. 127, pp. 87-142.
Améduri, B. and *Boutevin, B.*: Synthesis and Properties of Fluorinated Telechelic Monodispersed Compounds. Vol. 102, pp. 133-170.
Amselem, S. see Domb, A. J.: Vol. 107, pp. 93-142.
Andrady, A. L.: Wavelenght Sensitivity in Polymer Photodegradation. Vol. 128, pp. 47-94.
Andreis, M. and *Koenig, J. L.*: Application of Nitrogen-15 NMR to Polymers. Vol. 124, pp. 191-238.
Angiolini, L. see Carlini, C.: Vol. 123, pp. 127-214.
Anseth, K. S., Newman, S. M. and *Bowman, C. N.*: Polymeric Dental Composites: Properties and Reaction Behavior of Multimethacrylate Dental Restorations. Vol. 122, pp. 177-218.
Armitage, B. A. see O'Brien, D. F.: Vol. 126, pp. 53-58.
Arndt, M. see Kaminski, W.: Vol. 127, pp. 143-187.
Arnold Jr., F. E. and *Arnold, F. E.*: Rigid-Rod Polymers and Molecular Composites. Vol. 117, pp. 257-296.
Arshady, R.: Polymer Synthesis via Activated Esters: A New Dimension of Creativity in Macromolecular Chemistry. Vol. 111, pp. 1-42.

Bahar, I., Erman, B. and *Monnerie, L.*: Effect of Molecular Structure on Local Chain Dynamics: Analytical Approaches and Computational Methods. Vol. 116, pp. 145-206.
Ballauff, M. see Dingenouts, N.: Vol. 144, pp. 1-48.
Baltá-Calleja, F. J., González Arche, A., Ezquerra, T. A., Santa Cruz, C., Batallón, F., Frick, B. and *López Cabarcos, E.*: Structure and Properties of Ferroelectric Copolymers of Poly(vinylidene) Fluoride. Vol. 108, pp. 1-48.
Barshtein, G. R. and *Sabsai, O. Y.*: Compositions with Mineralorganic Fillers. Vol. 101, pp.1-28.
Batallán, F. see Baltá-Calleja, F. J.: Vol. 108, pp. 1-48.
Batog, A. E., Pet'ko, I. P., Penczek, P.: Aliphatic-Cycloaliphatic Epoxy Compounds and Polymers. Vol. 144, pp. 49-114.
Barton, J. see Hunkeler, D.: Vol. 112, pp. 115-134.
Bell, C. L. and *Peppas, N. A.*: Biomedical Membranes from Hydrogels and Interpolymer Complexes. Vol. 122, pp. 125-176.
Bellon-Maurel, A. see Calmon-Decriaud, A.: Vol. 135, pp. 207-226.
Bennett, D. E. see O'Brien, D. F.: Vol. 126, pp. 53-84.
Berry, G.C.: Static and Dynamic Light Scattering on Moderately Concentraded Solutions: Isotropic Solutions of Flexible and Rodlike Chains and Nematic Solutions of Rodlike Chains. Vol. 114, pp. 233-290.
Bershtein, V. A. and *Ryzhov, V. A.*: Far Infrared Spectroscopy of Polymers. Vol. 114, pp. 43-122.
Bigg, D. M.: Thermal Conductivity of Heterophase Polymer Compositions. Vol. 119, pp. 1-30.

Binder, K.: Phase Transitions in Polymer Blends and Block Copolymer Melts: Some Recent Developments. Vol. 112, pp. 115-134.
Binder, K.: Phase Transitions of Polymer Blends and Block Copolymer Melts in Thin Films. Vol. 138, pp. 1-90.
Bird, R. B. see *Curtiss, C. F.*: Vol. 125, pp. 1-102.
Biswas, M. and *Mukherjee, A.*: Synthesis and Evaluation of Metal-Containing Polymers. Vol. 115, pp. 89-124.
Bolze, J. see *Dingenouts, N.*: Vol. 144, pp. 1-48.
Boutevin, B. and *Robin, J. J.*: Synthesis and Properties of Fluorinated Diols. Vol. 102. pp. 105-132.
Boutevin, B. see *Amédouri, B.*: Vol. 102, pp. 133-170.
Boutevin, B. see *Améduri, B.*: Vol. 127, pp. 87-142.
Bowman, C. N. see *Anseth, K. S.*: Vol. 122, pp. 177-218.
Boyd, R. H.: Prediction of Polymer Crystal Structures and Properties. Vol. 116, pp. 1-26.
Briber, R. M. see *Hedrick, J. L.*: Vol. 141, pp. 1-44.
Bronnikov, S. V., Vettegren, V. I. and *Frenkel, S. Y.*: Kinetics of Deformation and Relaxation in Highly Oriented Polymers. Vol. 125, pp. 103-146.
Bruza, K. J. see *Kirchhoff, R. A.*: Vol. 117, pp. 1-66.
Burban, J. H. see *Cussler, E. L.*: Vol. 110, pp. 67-80.
Burchard, W.: Solution Properties of Branched Macromolecules. Vol. 143, pp. 113-194.

Calmon-Decriaud, A. Bellon-Maurel, V., Silvestre, F.: Standard Methods for Testing the Aerobic Biodegradation of Polymeric Materials. Vol 135, pp. 207-226.
Cameron, N. R. and *Sherrington, D. C.*: High Internal Phase Emulsions (HIPEs)-Structure, Properties and Use in Polymer Preparation. Vol. 126, pp. 163-214.
de la Campa, J. G. see *de Abajo, , J.*: Vol. 140, pp. 23-60.
Candau, F. see *Hunkeler, D.*: Vol. 112, pp. 115-134.
Canelas, D. A. and *DeSimone, J. M.*: Polymerizations in Liquid and Supercritical Carbon Dioxide. Vol. 133, pp. 103-140.
Capek, I.: Kinetics of the Free-Radical Emulsion Polymerization of Vinyl Chloride. Vol. 120, pp. 135-206.
Capek, I.: Radical Polymerization of Polyoxyethylene Macromonomers in Disperse Systems. Vol. 145, pp. 1-56.
Carlini, C. and *Angiolini, L.*: Polymers as Free Radical Photoinitiators. Vol. 123, pp. 127-214.
Carter, K. R. see *Hedrick, J. L.*: Vol. 141, pp. 1-44.
Casas-Vazquez, J. see *Jou, D.*: Vol. 120, pp. 207-266.
Chandrasekhar, V.: Polymer Solid Electrolytes: Synthesis and Structure. Vol 135, pp. 139-206
Charleux, B., Faust R.: Synthesis of Branched Polymers by Cationic Polymerization. Vol. 142, pp. 1-70.
Chen, P. see *Jaffe, M.*: Vol. 117, pp. 297-328.
Choe, E.-W. see *Jaffe, M.*: Vol. 117, pp. 297-328.
Chow, T. S.: Glassy State Relaxation and Deformation in Polymers. Vol. 103, pp. 149-190.
Chung, T.-S. see *Jaffe, M.*: Vol. 117, pp. 297-328.
Comanita, B. see *Roovers, J.*: Vol. 142, pp. 179-228.
Connell, J. W. see *Hergenrother, P. M.*: Vol. 117, pp. 67-110.
Criado-Sancho, M. see *Jou, D.*: Vol. 120, pp. 207-266.
Curro, J.G. see *Schweizer, K.S.*: Vol. 116, pp. 319-378.
Curtiss, C. F. and *Bird, R. B.*: Statistical Mechanics of Transport Phenomena: Polymeric Liquid Mixtures. Vol. 125, pp. 1-102.
Cussler, E. L., Wang, K. L. and *Burban, J. H.*: Hydrogels as Separation Agents. Vol. 110, pp. 67-80.

DeSimone, J. M. see *Canelas D. A.*: Vol. 133, pp. 103-140.
DiMari, S. see *Prokop, A.*: Vol. 136, pp. 1-52.
Dimonie, M. V. see *Hunkeler, D.*: Vol. 112, pp. 115-134.

Dingenouts, N., Bolze, J., Pötschke, D., Ballauf, M.: Analysis of Polymer Latexes by Small-Angle X-Ray Scattering. Vol. 144, pp. 1-48
Dodd, L. R. and *Theodorou, D. N.*: Atomistic Monte Carlo Simulation and Continuum Mean Field Theory of the Structure and Equation of State Properties of Alkane and Polymer Melts. Vol. 116, pp. 249-282.
Doelker, E.: Cellulose Derivatives. Vol. 107, pp. 199-266.
Dolden, J. G.: Calculation of a Mesogenic Index with Emphasis Upon LC-Polyimides. Vol. 141, pp. 189-245.
Domb, A. J., Amselem, S., Shah, J. and *Maniar, M.*: Polyanhydrides: Synthesis and Characterization. Vol. 107, pp. 93-142.
Dubrovskii, S. A. see Kazanskii, K. S.: Vol. 104, pp. 97-134.
Dunkin, I. R. see Steinke, J.: Vol. 123, pp. 81-126.
Dunson, D. L. see McGrath, J. E.: Vol. 140, pp. 61-106.

Economy, J. and *Goranov, K.*: Thermotropic Liquid Crystalline Polymers for High Performance Applications. Vol. 117, pp. 221-256.
Ediger, M. D. and *Adolf, D. B.*: Brownian Dynamics Simulations of Local Polymer Dynamics. Vol. 116, pp. 73-110.
Edwards, S. F. see Aharoni, S. M.: Vol. 118, pp. 1-231.
Endo, T. see Yagci, Y.: Vol. 127, pp. 59-86.
Erman, B. see Bahar, I.: Vol. 116, pp. 145-206.
Ewen, B, Richter, D.: Neutron Spin Echo Investigations on the Segmental Dynamics of Polymers in Melts, Networks and Solutions. Vol. 134, pp. 1-130.
Ezquerra, T. A. see Baltá-Calleja, F. J.: Vol. 108, pp. 1-48.

Faust, R. see Charleux, B: Vol. 142, pp. 1-70.
Fekete, E see Pukánszky, B: Vol. 139, pp. 109-154.
Fendler, J.H.: Membrane-Mimetic Approach to Advanced Materials. Vol. 113, pp. 1-209.
Fetters, L. J. see Xu, Z.: Vol. 120, pp. 1-50.
Förster, S. and *Schmidt, M.*: Polyelectrolytes in Solution. Vol. 120, pp. 51-134.
Freire, J. J.: Conformational Properties of Branched Polymers: Theory and Simulations. Vol. 143, pp. 35-112.
Frenkel, S. Y. see Bronnikov, S. V.: Vol. 125, pp. 103-146.
Frick, B. see Baltá-Calleja, F. J.: Vol. 108, pp. 1-48.
Fridman, M. L.: see Terent'eva, J. P.: Vol. 101, pp. 29-64.
Funke, W.: Microgels-Intramolecularly Crosslinked Macromolecules with a Globular Structure. Vol. 136, pp. 137-232.

Galina, H.: Mean-Field Kinetic Modeling of Polymerization: The Smoluchowski Coagulation Equation. Vol. 137, pp. 135-172.
Ganesh, K. see Kishore, K.: Vol. 121, pp. 81-122.
Gaw, K. O. and *Kakimoto, M.*: Polyimide-Epoxy Composites. Vol. 140, pp. 107-136.
Geckeler, K. E. see Rivas, B.: Vol. 102, pp. 171-188.
Geckeler, K. E.: Soluble Polymer Supports for Liquid-Phase Synthesis. Vol. 121, pp. 31-80.
Gehrke, S. H.: Synthesis, Equilibrium Swelling, Kinetics Permeability and Applications of Environmentally Responsive Gels. Vol. 110, pp. 81-144.
de Gennes, P.-G.: Flexible Polymers in Nanopores. Vol. 138, pp. 91-106.
Giannelis, E.P., Krishnamoorti, R., Manias, E.: Polymer-Silicate Nanocomposites: Model Systems for Confined Polymers and Polymer Brushes. Vol. 138, pp. 107-148.
Godovsky, D. Y.: Electron Behavior and Magnetic Properties Polymer-Nanocomposites. Vol. 119, pp. 79-122.
González Arche, A. see Baltá-Calleja, F. J.: Vol. 108, pp. 1-48.
Goranov, K. see Economy, J.: Vol. 117, pp. 221-256.
Gramain, P. see Améduri, B.: Vol. 127, pp. 87-142.

Grest, G.S.: Normal and Shear Forces Between Polymer Brushes. Vol. 138, pp. 149-184
Grosberg, A. and *Nechaev, S.*: Polymer Topology. Vol. 106, pp. 1-30.
Grubbs, R., Risse, W. and *Novac, B.*: The Development of Well-defined Catalysts for Ring-Opening Olefin Metathesis. Vol. 102, pp. 47-72.
van Gunsteren, W. F. see Gusev, A. A.: Vol. 116, pp. 207-248.
Gusev, A. A., Müller-Plathe, F., van Gunsteren, W. F. and *Suter, U. W.*: Dynamics of Small Molecules in Bulk Polymers. Vol. 116, pp. 207-248.
Guillot, J. see Hunkeler, D.: Vol. 112, pp. 115-134.
Guyot, A. and *Tauer, K.*: Reactive Surfactants in Emulsion Polymerization. Vol. 111, pp. 43-66.

Hadjichristidis, N., Pispas, S., Pitsikalis, M., Iatrou, H., Vlahos, C.: Asymmetric Star Polymers Synthesis and Properties. Vol. 142, pp. 71-128.
Hadjichristidis, N. see Xu, Z.: Vol. 120, pp. 1-50.
Hadjichristidis, N. see Pitsikalis, M.: Vol. 135, pp. 1-138.
Hall, H. K. see Penelle, J.: Vol. 102, pp. 73-104.
Hammouda, B.: SANS from Homogeneous Polymer Mixtures: A Unified Overview. Vol. 106, pp. 87-134.
Harada, A.: Design and Construction of Supramolecular Architectures Consisting of Cyclodextrins and Polymers. Vol. 133, pp. 141-192.
Haralson, M. A. see Prokop, A.: Vol. 136, pp. 1-52.
Hawker, C. J. see Hedrick, J. L.: Vol. 141, pp. 1-44.
Hedrick, J. L., Carter, K. R., Labadie, J. W., Miller, R. D., Volksen, W., Hawker, C. J., Yoon, D. Y., Russell, T. P., McGrath, J. E., Briber, R. M.: Nanoporous Polyimides. Vol. 141, pp. 1-44.
Hedrick, J. L. see Hergenrother, P. M.: Vol. 117, pp. 67-110.
Hedrick, J.L. see McGrath, J. E.: Vol. 140, pp. 61-106.
Heller, J.: Poly (Ortho Esters). Vol. 107, pp. 41-92.
Hemielec, A. A. see Hunkeler, D.: Vol. 112, pp. 115-134.
Hergenrother, P. M., Connell, J. W., Labadie, J. W. and *Hedrick, J. L.*: Poly(arylene ether)s Containing Heterocyclic Units. Vol. 117, pp. 67-110.
Hernández-Barajas, J. see Wandrey, C.: Vol. 145, pp. 123-182.
Hervet, H. see Léger, L.: Vol. 138, pp. 185-226.
Hiramatsu, N. see Matsushige, M.: Vol. 125, pp. 147-186.
Hirasa, O. see Suzuki, M.: Vol. 110, pp. 241-262.
Hirotsu, S.: Coexistence of Phases and the Nature of First-Order Transition in Poly-N-isopropylacrylamide Gels. Vol. 110, pp. 1-26.
Hornsby, P.: Rheology, Compoundind and Processing of Filled Thermoplastics. Vol. 139, pp. 155-216.
Hult, A., Johansson, M., Malmström, E.: Hyperbranched Polymers. Vol. 143, pp. 1-34.
Hunkeler, D., Candau, F., Pichot, C., Hemielec, A. E., Xie, T. Y., Barton, J., Vaskova, V., Guillot, J., Dimonie, M. V., Reichert, K. H.: Heterophase Polymerization: A Physical and Kinetic Comparision and Categorization. Vol. 112, pp. 115-134.
Hunkeler, D. see Prokop, A.: Vol. 136, pp. 1-52; 53-74.
Hunkeler, D. see Wandrey, C.: Vol. 145, pp. 123-182.

Iatrou, H. see Hadjichristidis, N.: Vol. 142, pp. 71-128
Ichikawa, T. see Yoshida, H.: Vol. 105, pp. 3-36.
Ihara, E. see Yasuda, H.: Vol. 133, pp. 53-102.
Ikada, Y. see Uyama, Y.: Vol. 137, pp. 1-40.
Ilavsky, M.: Effect on Phase Transition on Swelling and Mechanical Behavior of Synthetic Hydrogels. Vol. 109, pp. 173-206.
Imai, Y.: Rapid Synthesis of Polyimides from Nylon-Salt Monomers. Vol. 140, pp. 1-23.
Inomata, H. see Saito, S.: Vol. 106, pp. 207-232.
Irie, M.: Stimuli-Responsive Poly(N-isopropylacrylamide), Photo- and Chemical-Induced Phase Transitions. Vol. 110, pp. 49-66.
Ise, N. see Matsuoka, H.: Vol. 114, pp. 187-232.

Ito, K., Kawaguchi, S,: Poly(macronomers), Homo- and Copolymerization. Vol. 142, pp. 129-178.
Ivanov, A. E. see *Zubov, V. P.:* Vol. 104, pp. 135-176.

Jaffe, M., Chen, P., Choe, E.-W., Chung, T.-S. and *Makhija, S.:* High Performance Polymer Blends. Vol. 117, pp. 297-328.
Jancar, J.: Structure-Property Relationships in Thermoplastic Matrices. Vol. 139, pp. 1-66.
Johansson, M. see *Hult, A.:* Vol. 143, pp. 1-34.
Joos-Müller, B. see *Funke, W.:* Vol. 136, pp. 137-232.
Jou, D., Casas-Vazquez, J. and *Criado-Sancho, M.:* Thermodynamics of Polymer Solutions under Flow: Phase Separation and Polymer Degradation. Vol. 120, pp. 207-266.

Kaetsu, I.: Radiation Synthesis of Polymeric Materials for Biomedical and Biochemical Applications. Vol. 105, pp. 81-98.
Kakimoto, M. see *Gaw, K. O.:* Vol. 140, pp. 107-136.
Kaminski, W. and *Arndt, M.:* Metallocenes for Polymer Catalysis. Vol. 127, pp. 143-187.
Kammer, H. W., Kressler, H. and *Kummerloewe, C.:* Phase Behavior of Polymer Blends - Effects of Thermodynamics and Rheology. Vol. 106, pp. 31-86.
Kandyrin, L. B. and *Kuleznev, V. N.:* The Dependence of Viscosity on the Composition of Concentrated Dispersions and the Free Volume Concept of Disperse Systems. Vol. 103, pp. 103-148.
Kaneko, M. see *Ramaraj, R.:* Vol. 123, pp. 215-242.
Kang, E. T., Neoh, K. G. and *Tan, K. L.:* X-Ray Photoelectron Spectroscopic Studies of Electroactive Polymers. Vol. 106, pp. 135-190.
Kato, K. see *Uyama, Y.:* Vol. 137, pp. 1-40.
Kawaguchi, S. see *Ito, K.:* Vol. 142, p 129-178.
Kazanskii, K. S. and *Dubrovskii, S. A.:* Chemistry and Physics of „Agricultural" Hydrogels. Vol. 104, pp. 97-134.
Kennedy, J. P. see *Majoros, I.:* Vol. 112, pp. 1-113.
Khokhlov, A., Starodybtzev, S. and *Vasilevskaya, V.:* Conformational Transitions of Polymer Gels: Theory and Experiment. Vol. 109, pp. 121-172.
Kilian, H. G. and *Pieper, T.:* Packing of Chain Segments. A Method for Describing X-Ray Patterns of Crystalline, Liquid Crystalline and Non-Crystalline Polymers. Vol. 108, pp. 49-90.
Kishore, K. and *Ganesh, K.:* Polymers Containing Disulfide, Tetrasulfide, Diselenide and Ditelluride Linkages in the Main Chain. Vol. 121, pp. 81-122.
Kitamaru, R.: Phase Structure of Polyethylene and Other Crystalline Polymers by Solid-State ^{13}C/MNR. Vol. 137, pp 41-102.
Klier, J. see *Scranton, A. B.:* Vol. 122, pp. 1-54.
Kobayashi, S., Shoda, S. and *Uyama, H.:* Enzymatic Polymerization and Oligomerization. Vol. 121, pp. 1-30.
Koenig, J. L. see *Andreis, M.:* Vol. 124, pp. 191-238.
Kokufuta, E.: Novel Applications for Stimulus-Sensitive Polymer Gels in the Preparation of Functional Immobilized Biocatalysts. Vol. 110, pp. 157-178.
Konno, M. see *Saito, S.:* Vol. 109, pp. 207-232.
Kopecek, J. see *Putnam, D.:* Vol. 122, pp. 55-124.
Koßmehl, G. see *Schopf, G.:* Vol. 129, pp. 1-145.
Kressler, J. see *Kammer, H. W.:* Vol. 106, pp. 31-86.
Kricheldorf, H. R.: Liquid-Cristalline Polyimides. Vol. 141, pp. 83-188.
Krishnamoorti, R. see *Giannelis, E.P.:* Vol. 138, pp. 107-148.
Kirchhoff, R. A. and *Bruza, K. J.:* Polymers from Benzocyclobutenes. Vol. 117, pp. 1-66.
Kuchanov, S. I.: Modern Aspects of Quantitative Theory of Free-Radical Copolymerization. Vol. 103, pp. 1-102.
Kudaibergennow, S.E.: Recent Advances in Studying of Synthetic Polyampholytes in Solutions. Vol. 144, pp. 115-198.
Kuleznev, V. N. see *Kandyrin, L. B.:* Vol. 103, pp. 103-148.
Kulichkhin, S. G. see *Malkin, A. Y.:* Vol. 101, pp. 217-258.
Kummerloewe, C. see *Kammer, H. W.:* Vol. 106, pp. 31-86.

Kuznetsova, N. P. see Samsonov, G. V.: Vol. 104, pp. 1-50. Labadie, J. W. see Hergenrother, P. M.: Vol. 117, pp. 67-110.

Labadie, J. W. see Hedrick, J. L.: Vol. 141, pp. 1-44.
Lamparski, H. G. see O'Brien, D. F.: Vol. 126, pp. 53-84.
Laschewsky, A.: Molecular Concepts, Self-Organisation and Properties of Polysoaps. Vol. 124, pp. 1-86.
Laso, M. see Leontidis, E.: Vol. 116, pp. 283-318.
Lazár, M. and *RychlΩ, R.*: Oxidation of Hydrocarbon Polymers. Vol. 102, pp. 189-222.
Lechowicz, J. see Galina, H.: Vol. 137, pp. 135-172.
Léger, L., Raphaël, E., Hervet, H.: Surface-Anchored Polymer Chains: Their Role in Adhesion and Friction. Vol. 138, pp. 185-226.
Lenz, R. W.: Biodegradable Polymers. Vol. 107, pp. 1-40.
Leontidis, E., de Pablo, J. J., Laso, M. and *Suter, U. W.*: A Critical Evaluation of Novel Algorithms for the Off-Lattice Monte Carlo Simulation of Condensed Polymer Phases. Vol. 116, pp. 283-318.
Lesec, J. see Viovy, J.-L.: Vol. 114, pp. 1-42.
Liang, G. L. see Sumpter, B. G.: Vol. 116, pp. 27-72.
Lienert, K.-W.: Poly(ester-imide)s for Industrial Use. Vol. 141, pp. 45-82.
Lin, J. and *Sherrington, D. C.*: Recent Developments in the Synthesis, Thermostability and Liquid Crystal Properties of Aromatic Polyamides. Vol. 111, pp. 177-220.
López Cabarcos, E. see Baltá-Calleja, F. J.: Vol. 108, pp. 1-48.

Majoros, I., Nagy, A. and *Kennedy, J. P.*: Conventional and Living Carbocationic Polymerizations United. I. A Comprehensive Model and New Diagnostic Method to Probe the Mechanism of Homopolymerizations. Vol. 112, pp. 1-113.
Makhija, S. see Jaffe, M.: Vol. 117, pp. 297-328.
Malmström, E. see Hult, A.: Vol. 143, pp. 1-34.
Malkin, A. Y. and *Kulichkhin, S. G.*: Rheokinetics of Curing. Vol. 101, pp. 217-258.
Maniar, M. see Domb, A. J.: Vol. 107, pp. 93-142.
Manias, E., see Giannelis, E.P.: Vol. 138, pp. 107-148.
Mashima, K., Nakayama, Y. and *Nakamura, A.*: Recent Trends in Polymerization of a-Olefins Catalyzed by Organometallic Complexes of Early Transition Metals. Vol. 133, pp. 1-52.
Matsumoto, A.: Free-Radical Crosslinking Polymerization and Copolymerization of Multivinyl Compounds. Vol. 123, pp. 41-80.
Matsumoto, A. see Otsu, T.: Vol. 136, pp. 75-138.
Matsuoka, H. and *Ise, N.*: Small-Angle and Ultra-Small Angle Scattering Study of the Ordered Structure in Polyelectrolyte Solutions and Colloidal Dispersions. Vol. 114, pp. 187-232.
Matsushige, K., Hiramatsu, N. and *Okabe, H.*: Ultrasonic Spectroscopy for Polymeric Materials. Vol. 125, pp. 147-186.
Mattice, W. L. see Rehahn, M.: Vol. 131/132, pp. 1-475.
Mays, W. see Xu, Z.: Vol. 120, pp. 1-50.
Mays, J.W. see Pitsikalis, M.: Vol.135, pp. 1-138.
McGrath, J. E. see Hedrick, J. L.: Vol. 141, pp. 1-44.
McGrath, J. E., Dunson, D. L., Hedrick, J. L.: Synthesis and Characterization of Segmented Polyimide-Polyorganosiloxane Copolymers. Vol. 140, pp. 61-106.
McLeish, T.C. B., Milner, S. T.: Entangled Dynamics and Melt Flow of Branched Polymers. Vol. 143, pp. 195-256.
Mecham, S. J. see McGrath, J. E.: Vol. 140, pp. 61-106.
Mikos, A. G. see Thomson, R. C.: Vol. 122, pp. 245-274.
Milner, S. T. see McLeish, T. C. B.: Vol. 143, pp. 195-256.
Mison, P. and Sillion, B.: Thermosetting Oligomers Containing Maleimides and Nadiimides End-Groups. Vol. 140, pp. 137-180.
Miyasaka, K.: PVA-Iodine Complexes: Formation, Structure and Properties. Vol. 108. pp. 91-130.
Miller, R. D. see Hedrick, J. L.: Vol. 141, pp. 1-44.
Monnerie, L. see Bahar, I.: Vol. 116, pp. 145-206.

Morishima, Y.: Photoinduced Electron Transfer in Amphiphilic Polyelectrolyte Systems. Vol. 104, pp. 51-96.
Mours, M. see Winter, H. H.: Vol. 134, pp. 165-234.
Müllen, K. see Scherf, U.: Vol. 123, pp. 1-40.
Müller-Plathe, F. see Gusev, A. A.: Vol. 116, pp. 207-248.
Mukerherjee, A. see Biswas, M.: Vol. 115, pp. 89-124.
Mylnikov, V.: Photoconducting Polymers. Vol. 115, pp. 1-88.

Nagy, A. see Majoros, I.: Vol. 112, pp. 1-11.
Nakamura, A. see Mashima, K.: Vol. 133, pp. 1-52.
Nakayama, Y. see Mashima, K.: Vol. 133, pp. 1-52.
Narasinham, B., Peppas, N. A.: The Physics of Polymer Dissolution: Modeling Approaches and Experimental Behavior. Vol. 128, pp. 157-208.
Nechaev, S. see Grosberg, A.: Vol. 106, pp. 1-30.
Neoh, K. G. see Kang, E. T.: Vol. 106, pp. 135-190.
Newman, S. M. see Anseth, K. S.: Vol. 122, pp. 177-218.
Nijenhuis, K. te: Thermoreversible Networks. Vol. 130, pp. 1-252.
Noid, D. W. see Sumpter, B. G.: Vol. 116, pp. 27-72.
Novac, B. see Grubbs, R.: Vol. 102, pp. 47-72.
Novikov, V. V. see Privalko, V. P.: Vol. 119, pp. 31-78.

O'Brien, D. F., Armitage, B. A., Bennett, D. E. and *Lamparski, H. G.*: Polymerization and Domain Formation in Lipid Assemblies. Vol. 126, pp. 53-84.
Ogasawara, M.: Application of Pulse Radiolysis to the Study of Polymers and Polymerizations. Vol. 105, pp. 37-80.
Okabe, H. see Matsushige, K.: Vol. 125, pp. 147-186.
Okada, M.: Ring-Opening Polymerization of Bicyclic and Spiro Compounds. Reactivities and Polymerization Mechanisms. Vol. 102, pp. 1-46.
Okano, T.: Molecular Design of Temperature-Responsive Polymers as Intelligent Materials. Vol. 110, pp. 179-198.
Okay, O. see Funke, W.: Vol. 136, pp. 137-232.
Onuki, A.: Theory of Phase Transition in Polymer Gels. Vol. 109, pp. 63-120.
Osad'ko, I.S.: Selective Spectroscopy of Chromophore Doped Polymers and Glasses. Vol. 114, pp. 123-186.
Otsu, T., Matsumoto, A.: Controlled Synthesis of Polymers Using the Iniferter Technique: Developments in Living Radical Polymerization. Vol. 136, pp. 75-138.

de Pablo, J. J. see Leontidis, E.: Vol. 116, pp. 283-318.
Padias, A. B. see Penelle, J.: Vol. 102, pp. 73-104.
Pascault, J.-P. see Williams, R. J. J.: Vol. 128, pp. 95-156.
Pasch, H.: Analysis of Complex Polymers by Interaction Chromatography. Vol. 128, pp. 1-46.
Penczek, P. see Batog, A. E.: Vol. 144, pp. 49-114.
Penelle, J., Hall, H. K., Padias, A. B. and *Tanaka, H.*: Captodative Olefins in Polymer Chemistry. Vol. 102, pp. 73-104.
Peppas, N. A. see Bell, C. L.: Vol. 122, pp. 125-176.
Peppas, N. A. see Narasinhan, B.: Vol. 128, pp. 157-208.
Pet'ko, I. P. see Batog, A. E.: Vol. 144, pp. 49-114.
Pichot, C. see Hunkeler, D.: Vol. 112, pp. 115-134.
Pieper, T. see Kilian, H. G.: Vol. 108, pp. 49-90.
Pispas, S. see Pitsikalis, M.: Vol. 135, pp. 1-138.
Pispas, S. see Hadjichristidis: Vol. 142, pp. 71-128.
Pitsikalis, M., Pispas, S., Mays, J. W., Hadjichristidis, N.: Nonlinear Block Copolymer Architectures. Vol. 135, pp. 1-138.
Pitsikalis, M. see Hadjichristidis: Vol. 142, pp. 71-128.
Pötschke, D. see Dingenouts, N.: Vol 144, pp. 1-48.

Pospíšil, J.: Functionalized Oligomers and Polymers as Stabilizers for Conventional Polymers. Vol. 101, pp. 65-168.
Pospíšil, J.: Aromatic and Heterocyclic Amines in Polymer Stabilization. Vol. 124, pp. 87-190.
Powers, A. C. see Prokop, A.: Vol. 136, pp. 53-74.
Priddy, D. B.: Recent Advances in Styrene Polymerization. Vol. 111, pp. 67-114.
Priddy, D. B.: Thermal Discoloration Chemistry of Styrene-co-Acrylonitrile. Vol. 121, pp. 123-154.
Privalko, V. P. and Novikov, V. V.: Model Treatments of the Heat Conductivity of Heterogeneous Polymers. Vol. 119, pp 31-78.
Prokop, A., Hunkeler, D., Powers, A. C., Whitesell, R. R., Wang, T. G.: Water Soluble Polymers for Immunoisolation II: Evaluation of Multicomponent Microencapsulation Systems. Vol. 136, pp. 53-74.
Prokop, A., Hunkeler, D., DiMari, S., Haralson, M. A., Wang, T. G.: Water Soluble Polymers for Immunoisolation I: Complex Coacervation and Cytotoxicity. Vol. 136, pp. 1-52.
Pukánszky, B. and Fekete, E.: Adhesion and Surface Modification. Vol. 139, pp. 109-154.
Putnam, D. and Kopecek, J.: Polymer Conjugates with Anticancer Acitivity. Vol. 122, pp. 55- 124.

Ramaraj, R. and Kaneko, M.: Metal Complex in Polymer Membrane as a Model for Photosynthetic Oxygen Evolving Center. Vol. 123, pp. 215-242.
Rangarajan, B. see Scranton, A. B.: Vol. 122, pp. 1-54.
Raphaël, E. see Léger, L.: Vol. 138, pp. 185-226.
Reddinger, J. L. and Reynolds, J. R.: Molecular Engineering of π-Conjugated Polymers. Vol. 145, pp. 57-122.
Reichert, K. H. see Hunkeler, D.: Vol. 112, pp. 115-134.
Rehahn, M., Mattice, W. L., Suter, U. W.: Rotational Isomeric State Models in Macromolecular Systems. Vol. 131/132, pp. 1-475.
Reynolds, J.R. see Reddinger, J. L.: Vol. 145, pp. 57-122.
Richter, D. see Ewen, B.: Vol. 134, pp.1-130.
Risse, W. see Grubbs, R.: Vol. 102, pp. 47-72.
Rivas, B. L. and Geckeler, K. E.: Synthesis and Metal Complexation of Poly(ethyleneimine) and Derivatives. Vol. 102, pp. 171-188.
Robin, J. J. see Boutevin, B.: Vol. 102, pp. 105-132.
Roe, R.-J.: MD Simulation Study of Glass Transition and Short Time Dynamics in Polymer Liquids. Vol. 116, pp. 111-114.
Roovers, J., Comanita, B.: Dendrimers and Dendrimer-Polymer Hybrids. Vol. 142, pp 179-228.
Rothon, R. N.: Mineral Fillers in Thermoplastics: Filler Manufacture and Characterisation. Vol. 139, pp. 67-108.
Rozenberg, B. A. see Williams, R. J. J.: Vol. 128, pp. 95-156.
Ruckenstein, E.: Concentrated Emulsion Polymerization. Vol. 127, pp. 1-58.
Rusanov, A. L.: Novel Bis (Naphtalic Anhydrides) and Their Polyheteroarylenes with Improved Processability. Vol. 111, pp. 115-176.
Russel, T. P. see Hedrick, J. L.: Vol. 141, pp. 1-44.
Rychlý, J. see Lazár, M.: Vol. 102, pp. 189-222.
Ryzhov, V. A. see Bershtein, V. A.: Vol. 114, pp. 43-122.

Sabsai, O. Y. see Barshtein, G. R.: Vol. 101, pp. 1-28.
Saburov, V. V. see Zubov, V. P.: Vol. 104, pp. 135-176.
Saito, S., Konno, M. and Inomata, H.: Volume Phase Transition of N-Alkylacrylamide Gels. Vol. 109, pp. 207-232.
Samsonov, G. V. and Kuznetsova, N. P.: Crosslinked Polyelectrolytes in Biology. Vol. 104, pp. 1-50.
Santa Cruz, C. see Baltá-Calleja, F. J.: Vol. 108, pp. 1-48.
Sato, T. and Teramoto, A.: Concentrated Solutions of Liquid-Christalline Polymers. Vol. 126, pp. 85-162.
Scherf, U. and Müllen, K.: The Synthesis of Ladder Polymers. Vol. 123, pp. 1-40.
Schmidt, M. see Förster, S.: Vol. 120, pp. 51-134.
Schopf, G. and Koßmehl, G.: Polythiophenes - Electrically Conductive Polymers. Vol. 129, pp. 1-145.

Schweizer, K. S.: Prism Theory of the Structure, Thermodynamics, and Phase Transitions of Polymer Liquids and Alloys. Vol. 116, pp. 319-378.
Scranton, A. B., Rangarajan, B. and *Klier, J.*: Biomedical Applications of Polyelectrolytes. Vol. 122, pp. 1-54.
Sefton, M. V. and *Stevenson, W. T. K.*: Microencapsulation of Live Animal Cells Using Polycrylates. Vol. 107, pp. 143-198.
Shamanin, V. V.: Bases of the Axiomatic Theory of Addition Polymerization. Vol. 112, pp. 135-180.
Sherrington, D. C. see Cameron, N. R., Vol. 126, pp. 163-214.
Sherrington, D. C. see Lin, J.: Vol. 111, pp. 177-220.
Sherrington, D. C. see Steinke, J.: Vol. 123, pp. 81-126.
Shibayama, M. see Tanaka, T.: Vol. 109, pp. 1-62.
Shiga, T.: Deformation and Viscoelastic Behavior of Polymer Gels in Electric Fields. Vol. 134, pp. 131-164.
Shoda, S. see Kobayashi, S.: Vol. 121, pp. 1-30.
Siegel, R. A.: Hydrophobic Weak Polyelectrolyte Gels: Studies of Swelling Equilibria and Kinetics. Vol. 109, pp. 233-268.
Silvestre, F. see Calmon-Decriaud, A.: Vol. 207, pp. 207-226.
Sillion, B. see Mison, P.: Vol. 140, pp. 137-180.
Singh, R. P. see Sivaram, S.: Vol. 101, pp. 169-216.
Sivaram, S. and *Singh, R. P.*: Degradation and Stabilization of Ethylene-Propylene Copolymers and Their Blends: A Critical Review. Vol. 101, pp. 169-216.
Starodybtzev, S. see Khokhlov, A.: Vol. 109, pp. 121-172.
Steinke, J., Sherrington, D. C. and *Dunkin, I. R.*: Imprinting of Synthetic Polymers Using Molecular Templates. Vol. 123, pp. 81-126.
Stenzenberger, H. D.: Addition Polyimides. Vol. 117, pp. 165-220.
Stevenson, W. T. K. see Sefton, M. V.: Vol. 107, pp. 143-198.
Sumpter, B. G., Noid, D. W., Liang, G. L. and *Wunderlich, B.*: Atomistic Dynamics of Macromolecular Crystals. Vol. 116, pp. 27-72.
Suter, U. W. see Gusev, A. A.: Vol. 116, pp. 207-248.
Suter, U. W. see Leontidis, E.: Vol. 116, pp. 283-318.
Suter, U. W. see Rehahn, M.: Vol. 131/132, pp. 1-475.
Suzuki, A.: Phase Transition in Gels of Sub-Millimeter Size Induced by Interaction with Stimuli. Vol. 110, pp. 199-240.
Suzuki, A. and *Hirasa, O.*: An Approach to Artifical Muscle by Polymer Gels due to Micro-Phase Separation. Vol. 110, pp. 241-262.

Tagawa, S.: Radiation Effects on Ion Beams on Polymers. Vol. 105, pp. 99-116.
Tan, K. L. see Kang, E. T.: Vol. 106, pp. 135-190.
Tanaka, T. see Penelle, J.: Vol. 102, pp. 73-104.
Tanaka, H. and *Shibayama, M.*: Phase Transition and Related Phenomena of Polymer Gels. Vol. 109, pp. 1-62.
Tauer, K. see Guyot, A.: Vol. 111, pp. 43-66.
Teramoto, A. see Sato, T.: Vol. 126, pp. 85-162.
Terent'eva, J. P. and *Fridman, M. L.*: Compositions Based on Aminoresins. Vol. 101, pp. 29-64.
Theodorou, D. N. see Dodd, L. R.: Vol. 116, pp. 249-282.
Thomson, R. C., Wake, M. C., Yaszemski, M. J. and *Mikos, A. G.*: Biodegradable Polymer Scaffolds to Regenerate Organs. Vol. 122, pp. 245-274.
Tokita, M.: Friction Between Polymer Networks of Gels and Solvent. Vol. 110, pp. 27-48.
Tsuruta, T.: Contemporary Topics in Polymeric Materials for Biomedical Applications. Vol. 126, pp. 1-52.

Uyama, H. see Kobayashi, S.: Vol. 121, pp. 1-30.
Uyama, Y.: Surface Modification of Polymers by Grafting. Vol. 137, pp. 1-40.

Vasilevskaya, V. see Khokhlov, A.: Vol. 109, pp. 121-172.

Vaskova, V. see Hunkeler, D.: Vol.:112, pp. 115-134.
Verdugo, P.: Polymer Gel Phase Transition in Condensation-Decondensation of Secretory Products. Vol. 110, pp. 145-156.
Vettegren, V. I.: see Bronnikov, S. V.: Vol. 125, pp. 103-146.
Viovy, J.-L. and *Lesec, J.*: Separation of Macromolecules in Gels: Permeation Chromatography and Electrophoresis. Vol. 114, pp. 1-42.
Vlahos, C. see Hadjichristidis, N.: Vol. 142, pp. 71-128.
Volksen, W.: Condensation Polyimides: Synthesis, Solution Behavior, and Imidization Characteristics. Vol. 117, pp. 111-164.
Volksen, W. see Hedrick, J. L.: Vol. 141, pp. 1-44.

Wake, M. C. see Thomson, R. C.: Vol. 122, pp. 245-274.
Wandrey C., Hernández-Barajas, J. and *Hunkeler, D.*: Diallyldimethylammonium Chloride and its Polymers. Vol. 145, pp. 123-182.
Wang, K. L. see Cussler, E. L.: Vol. 110, pp. 67-80.
Wang, S.-Q.: Molecular Transitions and Dynamics at Polymer/Wall Interfaces: Origins of Flow Instabilities and Wall Slip. Vol. 138, pp. 227-276.
Wang, T. G. see Prokop, A.: Vol. 136, pp.1-52; 53-74.
Whitesell, R. R. see Prokop, A.: Vol. 136, pp. 53-74.
Williams, R. J. J., Rozenberg, B. A., Pascault, J.-P.: Reaction Induced Phase Separation in Modified Thermosetting Polymers. Vol. 128, pp. 95-156.
Winter, H. H., Mours, M.: Rheology of Polymers Near Liquid-Solid Transitions. Vol. 134, pp. 165-234.
Wu, C.: Laser Light Scattering Characterization of Special Intractable Macromolecules in Solution. Vol 137, pp. 103-134.
Wunderlich, B. see Sumpter, B. G.: Vol. 116, pp. 27-72.

Xie, T. Y. see Hunkeler, D.: Vol. 112, pp. 115-134.
Xu, Z., Hadjichristidis, N., Fetters, L. J. and *Mays, J. W.*: Structure/Chain-Flexibility Relationships of Polymers. Vol. 120, pp. 1-50.

Yagci, Y. and *Endo, T.*: N-Benzyl and N-Alkoxy Pyridium Salts as Thermal and Photochemical Initiators for Cationic Polymerization. Vol. 127, pp. 59-86.
Yannas, I. V.: Tissue Regeneration Templates Based on Collagen-Glycosaminoglycan Copolymers. Vol. 122, pp. 219-244.
Yamaoka, H.: Polymer Materials for Fusion Reactors. Vol. 105, pp. 117-144.
Yasuda, H. and *Ihara, E.*: Rare Earth Metal-Initiated Living Polymerizations of Polar and Nonpolar Monomers. Vol. 133, pp. 53-102.
Yaszemski, M. J. see Thomson, R. C.: Vol. 122, pp. 245-274.
Yoon, D. Y. see Hedrick, J. L.: Vol. 141, pp. 1-44.
Yoshida, H. and *Ichikawa, T.*: Electron Spin Studies of Free Radicals in Irradiated Polymers. Vol. 105, pp. 3-36.

Zubov, V. P., Ivanov, A. E. and *Saburov, V. V.*: Polymer-Coated Adsorbents for the Separation of Biopolymers and Particles. Vol. 104, pp. 135-176.

Subject Index

Activation energy of polymerization 30, 38
Activity coefficient 130, 152, 154
Actuators 60
Adhesives, conductive 60
ADMET 88
Agglomeration of particles 15, 50
Aggregation number 22, 23
Alkoxy PPV 91
Alkyl (C_1, tC_4) PEO macromonomer 34, 35
Alkyl spacer group 37
Alkylation 127, 128
Amphiphilic PEO macromonomer 6, 9, 10, 15
Applications 60
Aqueous polymerization 38
Average number of radicals per particle 15, 16

Band gap 64
Band theory 63
Batteries 60
Bi-unsaturated macromonomer 38
Bimolecular termination 27, 30, 35, 37, 38
Biosensors 60
Bipolaron 66
Bjerrum length 150
Block copolymers 20, 25-27
Branched structures 169
Brillouin zone 63

Chain branching 134
Chain transfer events 11, 42
Charge density 147, 161
Charge density parameter 150
Charge distance 155
Charge, effective 150
Cloud point 43
Coemulsifier 16, 17, 18, 25
Colloidal stability 42
Comb-like polymer 21
Compartmentalization 7, 8, 13, 38, 41, 50, 52
Complex formation 171

Concentration regimes 151
Condensed emulsifier layer 16
Conductor feedthroughs 60
Conductors, transparent 60
Contour length 150
Conventional surfactant 5, 13, 31, 46, 47, 50
Conversion curve 28, 30
Copolymerization 143, 145-148
Copolymers 109
Corrosion protection 60
Coulombic repulsion 5
Counterion activity 153-155
Counterion condensation 150
Coupling polymerization 70, 98
Critical concentration 151
Critical conversion 11
Critical flocculation concentration 47
Critical micellar concentration (CMC) 19, 23, 24, 26, 45
Crosslinker 48
Crosslinking 147, 148
Cyclopolymerization 132, 148

Debye length 151, 152
Dehydrohalogenation 87
Density 130, 138
Density gradient 169
Dewatering 174, 175
Diffusion controlled termination 35, 51
Diffusive degradation 17
Dodecyl PEO - macromonomer 30, 37
Doping 65
Drainage 173
Dry strength 174

Efficiency of stabilizer 32
Electrochromics 60, 112
Electroluminescence 60
Electropolymerization 68, 105, 107
Electrostatic repulsion 141, 146
Electrostatic stabilization 5, 49
Electrosteric stabilization 5, 49

Subject Index

Emeraldine 108
EMI shielding 60
Emulsifier 19
Emulsifier-free emulsion polymerization 15, 42, 45, 51
Emulsion polymerization 13, 45
Entry-rate coefficient of radicals 19
Equivalent conductivity 130, 131, 152, 157, 159, 161

Fibers, antistatic 60
Films, antistatic 60
First-order radical loss process 27, 38, 51, 52
Flocculation 173
Fuel cells 60
Functionality 44

Gel effect 8, 17, 28, 30, 33
Gel formation 39
Graft copolymer 25, 27, 30, 44, 52

Heck polymerization 93
HLB 21, 25, 30, 42, 43, 48, 52
Homogeneous nucleation 11, 15, 16, 27
Homopolymerization 135
HUFT theory 9, 15
Hydrodynamic radius 26
Hydrogels 48
Hydrogen bonds 42
Hydrophobic interaction 24

Initial rate 38
Initiation efficiency 11
Initiation of emulsion polymerization 14
Inks, conductive 60
Interface complex 17
Intermolecular interaction 36, 39
Interval I,II, and III 14, 15, 38
Intrinsic viscosity 165
Inverse emulsion polymerization 141
Inverse micelle 50
Inverse suspension polymerization 48
Ionic strength 165

Kinetic parameters 41
Knoevenagel polymerization 87
KPS initiated polymerization 34

LEDs 86, 87
Leucoemeraldine 108
Lewis acid polymerization 68, 97
Light scattering 165, 167
Liquid crystalline polymers 95

Mark-Kuhn-Houwink-Sakurada relationship 165, 166
Maximum rate of polymerization 42
Mechanism of dispersion polymerization 10
Mechanism of emulsion polymerization 14
Membranes 60
Methacryloyl-terminated PEO macromonomer 42
Micelization 26
Micelle 19, 20, 22
Microemulsion 17, 18
Miniemulsion 16
Model of dispersion polymerization 11, 12
Molar mass 143, 165
Molar mass distribution 168, 169
Molecular weight distribution 40
Monodisperse polymer particles 8, 11, 30, 31, 32
Monomer droplets 13, 18, 42
Muscle, artificial 60

Nigraniline 108
Nonionic emulsifier 20
Nonlinear optics 60
Nucleation mechanism 9, 45
Number of particles 14, 16, 29

Octadecyl PEO macromonomer 42
Oil-soluble initiator 13
Oligomeric radicals 14
Oligomers 110
Orbitals, molecular 62, 65
Organized association 28, 29, 42
Osmometry 167

Partial specific volume 167
Particle nucleation 39
Particle size 34
Particle size distribution (PSD) 49
PEDOT 97, 103
Peierl's instability 64
Penultimate model 146, 147
Pernigraniline 108
Phase separation 172
Pitch control 174
Polaron 66
Poly(alkylthiophenes) 99
Poly(hexadiyne) 73
Poly(p-phenylene) 77
-, hyperbranched 81
-, planarized 84
-, soluble 78
Poly(phenylene vinylene) 86
Polyacetylene 62, 63, 72

Polyaniline 106
Polydispersity 169
Polymeric steric stabilizer 9
Polymerizable group 44
Polymerization loci 7
Polymerization rate, overall 135
Polypyrroles 104
Polythiophene 67, 96
Precursor method 76, 89, 92, 110
Protoemeraldine 108
Pseudo-bulk kinetics 7

Quaternization 127, 128

Radical capture efficiency 42
Radical polymerization 136
Rate of dispersion polymerization 8, 27, 30
Rate of emulsion polymerization 15, 30
Rate of particle formation 17, 18, 19
Reaction medium 50
Reaction order 29, 33, 34, 51
Reactive surfactant 6, 13, 30, 52
Reactivity of macromonomer 33, 39, 42, 45
Reactivity ratios 145-147
Redox state 60
Refractive index 167
Regioregularity 101
Retention 173
Romp 70, 75, 77, 93

Second virial coefficient 166, 167
Seed emulsion polymerization 49
Self diffusion parameter 152
Sensors 112

Serum replacement 44
Siloles 111
Size exclusion chromatography (SEC) 168
Smith-Ewart kinetics 8, 14, 15
Soil conditioner 176
Solitons 65
Soluble precursors 82
Stabilizer 44
Steric stabilization mechanism 5, 21, 47, 49
Stille polymerization 60
Sulfobetains 128
Supercapacitors 60
Surface active oligomers 37
Suspension polymerization 48
Suzuki polymerization 72, 79, 84, 85, 98

Tensile strength 46
Textiles 60
Transition metal-containing polymers 111

Ultracentrifugation 166, 169
Unsaturated groups 13, 28, 38, 39

van der Waals interaction 5, 48
Viscosity 130, 138

Waste water treatment 175
Water-soluble polymers 74, 83, 91
Water-soluble initiator 13
Wet strength 173, 174
Wittig polymerization 86

Yamamoto polymerization 72, 98

Springer and the environment

At Springer we firmly believe that an international science publisher has a special obligation to the environment, and our corporate policies consistently reflect this conviction.

We also expect our business partners – paper mills, printers, packaging manufacturers, etc. – to commit themselves to using materials and production processes that do not harm the environment. The paper in this book is made from low- or no-chlorine pulp and is acid free, in conformance with international standards for paper permanency.

Printing: Saladruck, Berlin
Binding: Buchbinderei Lüderitz & Bauer, Berlin